Gender in STEM Education in the Arab Gulf Countries

Martina Dickson · Melissa McMinn · Dean Cairns
Editors

Gender in STEM Education in the Arab Gulf Countries

Springer

Editors
Martina Dickson
Emirates College for Advanced Education
Abu Dhabi, United Arab Emirates

Melissa McMinn
Open Polytechnic
Lower Hutt, New Zealand

Dean Cairns
Emirates College for Advanced Education
Abu Dhabi, United Arab Emirates

ISBN 978-981-19-9134-9 ISBN 978-981-19-9135-6 (eBook)
https://doi.org/10.1007/978-981-19-9135-6

This Springer imprint is published by the registered company Springer Nature Singapore Pte Ltd.
The registered company address is: 152 Beach Road, #21-01/04 Gateway East, Singapore 189721, Singapore

Foreword

The long-standing, persistent and widespread gender disparity in relation to STEM study is an important issue for STEM Educators. Despite numerous campaigns and initiatives, and changes to policy and practice, gender imbalances still stubbornly persist in most STEM fields of study, across age ranges and across most countries and regions. If we look at Higher Education, the sector with which I am most familiar, we can see that globally approximately a third of students enrolled in STEM study are women (UNESCO, 2017). Gender seems to impact a subject choice even within the STEM umbrella—for example, the global enrolment in ICT study by women is 3%, and in the natural sciences, mathematics and statistics it is 5% (ibid.).

Unsurprisingly, and as the chapters in this book expertly discuss, there has been much research and scholarship attempting to investigate the reasons behind such stubbornly persistent gendered patterns. As many feminists have argued, there seems to be a widespread social and cultural association of STEM fields with the masculine—an association that influences students' own choices as to gender 'appropriate' pathways; students' perceptions of their own abilities and expertise in STEM; and crucially influences the ways in which teachers, parents and mentors perceive students' ability and potential 'fit' into STEM fields, and thus the encouragement and support given to students in pursuing such lines of study.

Nevertheless, there are also intriguing regional and national differences within this overall picture. For example, women comprise 16% of natural sciences, mathematics and statistics students in Côte D'Ivoire HE institutions—whereas in Bahrain the figure is 86% (UNESCO, 2017). These wide differences speak to the importance of understanding the social and cultural contexts of STEM Education in relation to gender—as well as emphasising that gendered variations cannot be understood as the result of 'natural' biological differences between two neatly categorised sexes—the picture is a lot more complex.

As researchers we also need to do delve beyond the sometimes over-simplistic numerical statistics regarding gender and STEM study, to explore in-depth the many facets of influence and experience that contextualise students' journeys into, within and beyond STEM study (or the reasons why this journey is not undertaken). The chapters in this volume are an excellent example of such scholarship, exploring the gendered dynamics of STEM education across different national contexts within the Gulf region. The chapters employ a wide range of methodological tools to tease out some of the intricate ways in which students' and teachers' perceptions are complexly gendered, and the implications of this for finding ways to push through and beyond persistent gender disparities in STEM in Gulf countries and beyond.

A number of chapter authors note that their research has been undertaken against a backdrop of government investment and prioritisation of a knowledge-based economy, and the role that STEM plays within this. This echoes the importance given by national governments across the globe in investing in STEM—for reasons of economic gain, but also at least partly in response to the role STEM and other disciplinary fields can play in combating key global difficulties such as the current and potential future pandemics, and the potentially devastating impact of climate change. To combat these challenges we need to know we are drawing on the best possible pool of students and professionals educated in STEM, those who have embarked in STEM because they—and others—have seen it as an appropriate choice regardless of gender (and/or other aspects of identity such as 'race'/ethnicity, socio-economic background, age, or disability). It is a national and international concern if potential students are discouraged from paths they would otherwise have taken. Indeed, it is a matter of human rights for all individuals to be able to choose access to whatever fields of education and employment they feel they can flourish in, speaking directly to UN Sustainable Development Goals 4 and 5 on Quality Education and Gender Equality, as well as contributing to SDG 8 on Decent Work and Economic Growth.

The scholarship in this book contributes to meeting these goals, covering a wide range of sectors and foci, from elementary middle and high school students to academic staff and STEM leaders—and across a range of contexts in this fascinating region. The research in this book has much to offer academics, policymakers and practitioners not only within the Gulf and MENA regions but also further afield. At the time of writing the world continues to struggle with the Covid-19 pandemic, highlighting the continued importance of ensuring equality and the greatest possible diversity in our STEM fields across the globe. With its focus on implications for education policy and practice, the chapters in this book are an important and welcome contribution to our international knowledge in this area and a tool with which to implement sustainable change.

March 2022

Dr. Barbara Read
Reader in Gender and Social Inequalities
School of Education
University of Glasgow
Glasgow, Scotland

Preface

STEM (science, technology, engineering and mathematics) subjects are at the top of most countries' national agendas in terms of planning for future economies, investment in knowledge and development, industries, commerce, and innovative growth. As each country grapples with the logistics around increasing the key performance indicators (KPIs) related to STEM growth, each has at its foundation the critical need for a future projection of a cadre of highly qualified, skilled human capital to work in these industries in a sustainable way. Being able to offer this pool relies on having a steady stream of motivated and competent graduates from STEM degrees in tertiary institutions, and this in turn relies upon a steady flow of students from secondary schools who make the decision to follow those pathways. Attitudes towards STEM subjects are often formed early on, and for that reason, how students think and feel about learning these subjects at both the primary and secondary educational level is important. Not to say that these attitudes and efficacies are totally unmalleable, but they are often powerful indicators of future behaviours, and so STEM attitudes and learning experiences are important at each stage of the educational journey. We also know that gender imbalance and inequities in representation persist at each stage of this educational journey, from attitudes and efficacies, to equitable representation later in the STEM workplace.

The motivation for compiling and writing this book, came about as a result of many hours of discussion among the editorial team about a variety of aspects of STEM, and through collaborative research based both on empirical and secondary data projects. The commonality though was that as researchers in our current geographical location (the United Arab Emirates, positioned in the middle of the Arabian Gulf) we often experienced a lack of research specifically looking at gender issues in STEM education in this region. While many of the patterns and challenges faced in the Gulf region are not dissimilar to those faced by countries in other parts of the world, the particular nuances of culture, geographical context, history, and traditions also has an important role to play in the way in which STEM is taught, received, and perpetuated. So, the idea of creating our own book to fill this literature chasm, developed from there.

Our book begins with three chapters that focus upon STEM beliefs and identity, given what is already well known and understood in the research literature about the critical effect of these on a number of factors, including student achievement and science career aspiration. Chapter 1 discusses ways in which STEM subjects have historically been perceived as being 'boys' subjects in many contexts, and presents data which indicates that in the GCC region, these attitudes (such as confidence in and value of, science) may be changing. Stereotypical beliefs that science and other STEM-related subjects are for boys, based on an out-dated, but persistent, notion that boys are innately better than girls at these subjects, are discussed. This leads into Chap. 2, which examines scientific epistemological beliefs—the nature and acquisition of science knowledge; how these beliefs impact males and females, their associated beliefs about learning, and their science achievement. Chapter 2 ends with a section on what the implications of these findings might be to classroom practice and methodologies. As the final chapter in this first section of the book, Chapter 3 showcases empirical data of drawings of scientists using the methodology of the famous 'DAST'—Draw a Scientist Test—but specifically analyses drawings of female scientists which girls in the United Arab Emirates have drawn. The chapter presents data on the differences between the genders of the scientists which boys and girls tend to draw, and then focuses in on the specific characteristics of scientists which girls draw. Implicit in this analysis is at last a part-assumption that the drawings may indicate perceptions of not only what scientists look like, the kinds of work they do, the characteristics they appear to have, but also *who* scientists tend to be.

So, the first section of our book delves into science beliefs and identity, and possible implications of such beliefs and identity to the ways in which students interact with science as a school subject, and consequently consider science as a career possibility. This then flows into the second section of the book, where attitudes towards, and understanding of, two other components of STEM, mathematics and technology, are explored. This section considers a variety of students' outcomes that can impact, and be impacted by, STEM learning. Chapters 4 and 5 examine attitudes and anxiety, respectively, towards mathematics, and considers the relationship of these to the learning environment. Chapter 4 explores the possibly altering perceptions that middle school girls in the UAE may have towards learning mathematics, and Chap. 5 follows on from this with an in-depth exploration of mathematics anxiety in females—the causes and extent of this, and what practices, strategies and policies could potentially be put into place to break this cycle of anxiety leading to non-engagement. Chapter 6 investigates the motivational and behavioural characteristics important for ICT literacy. The final chapter in this section (Chapter 7) explores a critical issue of our day—whether the future adult citizens of the Gulf countries are well-informed of environmental issues, and whether a gender difference in understanding of environmental issues in the Gulf exists.

In the third section of the book, the two final chapters look at the experiences and representation of women in STEM educational phases, both as receivers/participants and as providers of STEM education. Chapter 8 charts the narrative journeys of female leaders in STEM fields, looking specifically at their educational journeys through school and university, their experiences, influences and factors leading to their positions of success. Finally, Chapter 9 culminates the book's narrative on all of the factors which could potentially lead to gendered imbalance within STEM fields in the Gulf region. It presents analysis of staffing data in STEM departments in higher education institutions in the UAE, which also serves as a microcosmic reflection of the status in the Gulf region, and uses this to draw out possible explanations for the relatively low representation of women in these fields, and outlines the possible consequences of this to other women in the field.

It is hoped that the reader will benefit from the holistic nature of this book, which encompasses gendered differences and experiences across a wide variety of educational sectors, STEM components, and contexts.

Abu Dhabi, United Arab Emirates — Martina Dickson
Lower Hutt, New Zealand — Melissa McMinn
Abu Dhabi, United Arab Emirates — Dean Cairns

Contents

Part I STEM Beliefs and Identity

1 'Science is a Boys' Subject'—Changing Perceptions in the Arabian Gulf ... 3
Melissa McMinn

2 Epistemological Beliefs About Science and Their Relations to Gender, Attitudes to Science and Science Achievement in UAE Schools ... 31
Dean Cairns

3 The Drawing a Scientist Test (DAST): How Do Girls in the UAE Present Visual Characteristics of Female Scientists, and What Does This Mean for Gender Equity of Science Careers? ... 61
Martina Dickson and Melissa McMinn

Part II Attitudes and Understanding in STEM

4 Changing Perceptions of the Learning Environment and Attitudes Towards Mathematics Through Inquiry-Based Learning: Girls in Middle School Classrooms in the UAE ... 85
Jennifer Robinson and Jill Aldridge

5 Mathematics Anxiety in Females—Breaking the Cycle ... 119
Melissa McMinn

6 Gender Differences in ICT Literacy: ICT-Related Individual Characteristics and Enabling Factors ... 153
Ieda M. Santos and Shaljan Areepattamannil

7 Understanding of Environmental Issues Across Two Gulf Countries: Do Girls Know More Than Boys in UAE Schools? ... 173
Dean Cairns

Part III Gendered Representation and Experiences in STEM Education

8 Female STEM Leadership in the Gulf: Journeys Through Education 199
Martina Dickson and Masada Al Harthi

9 Gender Representation in STEM Departments in Higher Education Institutions in the UAE 221
Martina Dickson and Masada Al Harthi

Editors and Contributors

About the Editors

Dr. Martina Dickson holds a Ph.D. in Physics from the University of London and MA in Gender, education and International Development from the IoE (UCL) in London. She has held a variety of teaching and advisory positions in Greece, Oman, Hong Kong and the UAE and is currently a professor in the Curriculum and Instruction Department at Emirates College for Advanced Education in the UAE. Her research interests include gender in education and science pedagogy, and she has led a number of research projects over the last decade, including studies of academic parenthood, female STEM leadership, gendered choices and perceptions of scientists and science careers.

Dr. Melissa McMinn has worked in in-service and pre-service teacher education for 15 years and in postgraduate education for eight years. She is currently facilitating a new suite of initial education teacher programmes for the Open Polytechnic in New Zealand. She holds a Master of Education and a Doctor of Philosophy in Mathematics and Science Education and has achieved the status of the senior fellow (SFHEA) in recognition of her teaching and learning support in higher education. She is an active researcher and over the past decade has led projects in mathematics and science anxiety in university students and teachers, learning environments research and co-researched a number of projects including pedagogy in higher education and children's perceptions of science and technology, among others.

Dr. Dean Cairns holds a Ph.D. in Chemistry from the University of Sussex and is a doctoral candidate for an Ed.D. in Education at Bath University. He has worked as both a research scientist and a science educator. He is currently an associate professor in Curriculum and Instruction at the Emirates College for Advanced Education. He has published widely in high-quality international journals in the fields of polymer chemistry and science education. Recent publications have included investigating the impact of inquiry-learning approaches on science outcomes and students' perceptions of using technology during inquiry-related activities. His current research focuses on students' epistemic beliefs about science and their relationships with science conceptual knowledge development.

Contributors

Masada Al Harthi Emirates College for Advanced Education, Abu Dhabi, UAE

Jill Aldridge Curtin University, Perth, Australia

Shaljan Areepattamannil Emirates College for Advanced Education, Abu Dhabi, UAE

Dean Cairns Emirates College for Advanced Education, Abu Dhabi, UAE

Martina Dickson Emirates College for Advanced Education, Abu Dhabi, UAE

Melissa McMinn Open Polytechnic, Lower Hutt, New Zealand

Jennifer Robinson Emirates College for Advanced Education, Abu Dhabi, UAE

Ieda M. Santos Emirates College for Advanced Education, Abu Dhabi, UAE

Peer Reviewers, in addition to Editorial Team

Dr. Nagla Ali Emirates College for Advanced Education, Abu Dhabi, UAE

Dr. Laila Boisselle UWI-ROYTEC, Trinidad and Tobago

Dr. Jennifer Robinson Emirates College for Advanced Education, Abu Dhabi, UAE

Dr. Andres Sandoval-Hernandez University of Bath, England, UK

Part I
STEM Beliefs and Identity

Beliefs about STEM are thought to be the personal philosophies, assumptions and judgements about the nature of STEM subjects, as well as about the teaching and learning of them, and are part of the foundation upon which behaviours are built. Such beliefs have been observed to be subject specific and can range between 'naïve' and 'sophisticated'. For example, in mathematics, this is the difference between believing knowledge is certain, rigid and handed down by authority, and believing knowledge is tentative, complex, creative and derived from reason.

Science identity relates to how we see ourselves in relation to scientific activities. It includes who we think we are, what we believe we are capable of doing, and what we aspire to do and become as it relates to science. A student's science identity is based on their perceptions of themselves and how they perceive others view them as they engage in science-related tasks. If students hold beliefs that scientists are a different kind of person than themselves, those students might conclude they are not a 'science person'. Such beliefs may include epistemological beliefs about what science is, and beliefs around who can do well in science and who scientists are. A discrepancy between a student's personal sense of identity and a science identity can hamper persistence in STEM, therefore a healthy science identity, based on sophisticated beliefs, is essential to learning. The following section examines a variety of beliefs, and how these may impact science identity, science literacy and other outcomes that may determine how students engage in science, and for how long.

Chapter 1
'Science is a Boys' Subject'—Changing Perceptions in the Arabian Gulf

Melissa McMinn

Abstract In the United Arab Emirates, and internationally, great emphasis has been placed upon improving students' performance in science, technology, engineering and mathematics (STEM) subjects and encouraging STEM careers. However, international studies have shown that even where there are no significant differences in mathematics and science achievement, women still remain less likely to enrol in STEM fields in higher education. Gender stereotyping, self-efficacy and attitudes towards science can potentially influence young women's (and men's) aspirations to pursue studies and careers in these fields. The sciences have a long association with 'masculinity', and evidence suggests that many children perceive that science, particularly the physical sciences, are 'for boys' and that scientists are generally male. However, attitudes towards science appear to be changing, with both male and female students in the Gulf Cooperation Council (GCC) region responding favourably to survey questions about liking, feeling confident in and valuing, science. Current strategies to overcome stereotypes; boost STEM self-efficacy, particularly for female students; positively utilise the influence of teachers; and use role models and other supports are discussed, and recommendations for what still can be done are made.

Keywords STEM · Stereotypes · Masculinity · Self-efficacy · Gender

Introduction

Research such as that in *The Psychology of Sex Differences* (Maccoby & Jacklin, 1974) has taught us, with long-lasting effect, that boys are just better at mathematics and science than girls, due to innate differences in the genders. Although the book actually reported a *lack* of gender differences in areas in which common stereotypes and prior studies claimed existed, the 'science is for boys' mantra has permeated.

M. McMinn (✉)
Open Polytechnic, Lower Hutt, New Zealand
e-mail: Melissa.McMinn@openpolytechnic.ac.nz

M. Dickson et al. (eds.), *Gender in STEM Education in the Arab Gulf Countries*,
https://doi.org/10.1007/978-981-19-9135-6_1

Maccoby and Jacklin (1974) *did* conclude that, on average, by high school, boys outperformed girls in quantitative and spatial abilities, while girls demonstrated superior verbal ability over boys. These findings were later corroborated by many other studies, including a meta-analysis conducted by Hedges and Nowell in the U.S.A. (1995), and a cross-national study of 13-year-old children in the United States and Thailand (Engelhard, 1990). By the early 1990s, studies were showing that gender effect sizes for science were somewhat larger than those for mathematics (see, for example, Cleary, 1991; Hedges & Nowell, 1995; Linn, 1991). Findings indicated that boys outperformed girls on science tests across all age groups (Cleary, 1991), and that the 'male advantage' on science tests was up to 0.50 standard deviations (Hedges & Nowell, 1995). This advantage was also shown repeatedly to increase with age and was more pronounced in the 'hard' sciences (Beller & Gafni, 1996; Friedler & Tamir, 1990; Sjoberg, 1988).

However, studies such as these largely focused on achievement, usually measured by written tests, and without consideration of the many variables that may impact achievement differently across genders. This chapter discusses some of these variables, including past research and traditional societal positions, stereotyping, self-efficacy and attitudes towards science, and how these may impact both achievement and aspirations for further study and careers in the sciences. The current state of the gender gap is discussed, with an emphasis on how the countries of the Gulf Cooperation Council (GCC) are faring. Areas to consider in relation to stubborn stereotypes around science, boosting the self-efficacy of girls, the influence of teachers and how girls can be best supported are discussed here, with recommendations for future strategies.

Research and Societal Positions

Prior to the 1970s, academic differences between the genders, usually referred to as *sex differences,* were believed to be biological, and thus genetically determined (Bell & Norwood, 2007). These differences were seen to be fixed, and therefore, were accepted, and even expected. Such a belief absolved schools and teachers from working to change them. Around the 1970s, researchers began to favour the term *sex-related*, which implied "that while the behaviour of concern was clearly related to the sex of the subjects, it was not necessarily genetically determined" (Fennema, 2000, p. 2). Near the end of the decade, an influential paper, *Toward a Redefinition of Sex and Gender*, asserted that the use of the term *gender* "serves to reduce assumed parallels between biological and psychological sex or at least to make explicit any assumptions of such parallels" (Unger, 1979, p. 1,086). Since then, with increasing prevalence, researchers have used the term *gender differences* to encompass biological as well as social and environmental factors when conceptualising the achievement differences between males and females.

Meta-analyses have shown that the gender gap reported in studies has declined steeply over the years (Feingold, 1994), and that reports of gender difference in achievement after 1974 were half what they were before (Hyde et al., 1990). This was largely attributed to the changing attitudes about which careers are more appropriate for girls and which are more appropriate for boys.

Despite these improvements, the 'science is for boys' message is still prevalent. Societal messages often reflect people's beliefs and attitudes about academic subjects and who is likely to be good at them. These beliefs spread throughout cultures through mass media, books, parents, teachers and school administrators, and can perpetrate and reinforce stereotypes. For example, male STEM characters (62.9%) significantly outnumbered female STEM characters (37.1%) in film, television, and streaming content in the decade 2007–2017 (The Lyda Hill Foundation & Geena Davis Institute on Gender in Media, 2018). Scientists, specifically, are more often represented as males in TV formats than females (see, for example, Long et al., 2010; Steinke & Tavarez, 2018). Even in science journals that routinely publish analyses and general interest articles about gender and STEM, such as *Science* and *Nature*, unequal proportions of male and female representation in authorship, content, and visibility in photographs were found, significantly favouring males (Loverock & Hart, 2018). This led the authors to state that "science journals provide a model of the archetypal scientist—if that archetype is predominantly masculine, then the scientific culture itself may be contributing to gender disparity" (p. 755). In online science education resources aimed at primary school students, a visual content analysis showed that boys were depicted more often that girls, and men more often than women (Kerkhoven et al., 2016). Furthermore, more men than women were depicted in science professions in these resources, and more women than men were depicted as teachers. Even popular toys can impact beliefs. The now infamous Teen Talk Barbie, an electronic talking version of the Mattel doll, manufactured in 1992, was programmed to say, 'Math class is hard'. In a press conference, a Mattel spokesperson defended the doll and claimed that Teen Talk Barbie was a wonderful role model for today's girls; the perfect nineties girl (Bell & Norwood, 2007). The subsequent backlash caused Mattel to withdraw the phrase and conceded that they had not fully considered the negative implications. Such examples demonstrate the pervasive nature of the 'science is for boys' message and demonstrate the difficult undertaking for gender equity in STEM-related activities.

Stereotyping and Masculinity

> The male stereotype of science and of a scientist is persistent and appears as early as in kindergarten, while the association of science with men is especially persistent among older children (Makarova et al., 2019, p. 3).

Indeed, studies that used the Draw a Scientist Test (DAST) method reported that students from kindergarten to high school have largely perceived scientists as males. In the original DAST study involving 4,807 kindergarten to grade five students, only 28 pictures depicted a female—all drawn by girls (Chambers, 1983). In a more recent meta-analysis based on 78 U.S. DAST studies (N = 20,860) among grade K-12 children across several decades, an increase in children's depictions of female scientists was shown in later decades (Miller et al., 2018). However, female scientists appeared only in drawings by young children, with older children still relating science to men (Miller et al., 2018). The authors concluded that despite the increase of women's representation in science over the last decades, children still observe more male than female scientists in their social environments. (See Chap. 3 for results of a DAST study conducted in the UAE).

Research on gender stereotypes has revealed that science is not only associated with a male person but that masculine traits are also attributed to it. From an early age, children show strong gender-typing of mathematics and science as inherently masculine, while reading and language are regarded as feminine (Fennema et al., 1990; Halim & Ruble, 2010). This seems to occur even though young children do not have a clear understanding of science subjects (Archer et al., 2010). These associations generally endorse beliefs that males are naturally more talented and interested in mathematics and science than females (Cvencek et al., 2015; Farrell & McHugh, 2017; Smyth & Nosek, 2015; Steele, 2003; Steffens & Jelenec, 2011). As negative stereotypes have been found to impair girls' performance on standardised tests through the mechanism of stereotype threat (Spencer et al., 1999; Steele, 1997), it is easy to see how a negative cycle can be perpetuated. Indeed, national differences in the strength of gender-science stereotypes are correlated with national gender gaps in science achievement scores (Nosek et al., 2009). Even when such stereotypes are not explicitly endorsed, implicit gender stereotypes persist into adulthood and are found across cultures (Nosek et al., 2002).

As dichotomous constructs, gender stereotypes infer that what is masculine is not feminine and vice versa (Deaux & LaFrance, 1998; Renfrow & Howard, 2013; Worell, 2001). Where science subjects were rated as masculine by school and university students, they were also associated with a set of attributes commonly associated with masculinity, such as being hard, complex and based on thinking rather than on feelings (Weinreich-Haste, 1981). Similarly, a study among secondary school students in Switzerland showed that semantic traits associated with mathematics and physics included adjectives such as hard, serious, distant, sober, strict, robust and rigid, whereas the female gender was strongly associated with traits such as being soft, playful, soulful, dreamy, lenient, frail and flexible (Makarova & Herzog, 2015). Where certain characteristics are perceived to be important for succeeding in science, and those characteristics are more commonly ascribed to males, gender role biases are formed. Gender role biases towards STEM disciplines have been empirically shown to exist and can be held by parents, teachers and employers (Carli et al., 2016; Hand et al., 2017; Reilly et al., 2019). This may explain why several studies over the past decades have demonstrated that boys were far more likely than girls to have been exposed to scientific activities, such as using microscopes, telescopes, barometers and electricity meters, and enrolled in extracurricular science courses (Dimitrov, 1999; Linn, 1985; Martinez & Mead, 1988; NAEP, 1988; OECD, 2016; Sobieraj & Krämer, 2019).

Interestingly, gender-typing across science branches is not equal and has fluctuated over time. A study on the gendered perceptions of different academic subjects involving school and university students reported that mathematics, physics and chemistry had the strongest masculine connotations (Weinreich-Haste, 1981). In contrast, a study on gendered perceptions of school subjects among 11–12-year-old students, reported that while physics was rated as significantly more masculine, chemistry and mathematics were rated as neither masculine nor feminine (Archer & MacRae, 1991). In 2019, Makarova et al. found that mathematics was most strongly perceived as a masculine subject by both female and male secondary school students, followed by physics, and chemistry had the weakest masculine connotations of the three subjects studied. Female adolescents have been found to perceive all three subjects considerably more strongly as a male domain than do male students (Kessels et al., 2006; Makarova et al., 2019; Nosek et al., 2002; Steffens et al., 2010).

When the gender stereotypes of physics held by high school students were analysed, being interested in physics was found to be associated with the male gender (Kessels, 2005; Kessels et al., 2006), and for girls, being interested in physics endangered their self-identification with the female gender (Kessels et al., 2006). High school students also attributed better performance in STEM subjects to boys (Hand et al., 2017), and first-year male university students held negative stereotypes of women's engineering and mathematical ability (Jones et al., 2013). Furthermore, even women who had selected mathematics-intensive university majors had difficulties in associating their major with themselves because they associated mathematics with the male gender (Nosek et al., 2002).

Perhaps fuelled by stereotypical beliefs, are possible misperceptions that surround STEM careers that may deter their uptake. Females' perceptions of STEM careers are generally stereotyped as being thing-oriented (the preference for working with objects), and not people-orientated (the preference for working with or helping other people), typically things that are unappealing to women (Kang et al., 2019; Konrad et al., 2000). In a meta-analysis conducted to explain gender disparities across STEM fields, Su and Rounds (2015) found engineering disciplines, representing things-orientated working environments, were significantly favoured by men and in contrast, social sciences and medical services, representing people-orientated working environments, were largely favoured by women. The perception that STEM careers are antithetical to communal goals significantly impedes women's STEM career trajectories since people-orientated opportunities are highly valued when women make vocational decisions. (Diekman et al., 2010; Kang et al., 2019). STEM careers, in fact, hold the key to helping people and society (Kang et al., 2019). This represents a mismatch in what women want from a career, and what a STEM career is (incorrectly) perceived to be able to give them. Interestingly, Brown et al. (2015) reported that regardless of university students' major or gender, when they had the chance to perceive STEM careers as supporting communal values, their interest in STEM careers increased.

Contributing Factors

Given the pervasive stereotypes and masculine imagery associated with science and propagated throughout society, it is not surprising that many studies have shown that males hold more positive self-efficacy, self-images and attitudes in relation to the topic.

Self-efficacy

Women's and girls' self-efficacies can be undermined if the common negative stereotypes associated with women in science and mathematics are truly believed (Reilly et al., 2019; Schmader et al., 2004; Stout et al., 2013), and many researchers have indicated that a lack of self-efficacy in STEM disciplines contributes to the disparity between men and women pursuing careers in these fields (see, for example, Larose et al., 2006; Rittmayer & Beier, 2009; Zeldin & Pajares, 2000). For example, it has been noted that "…women aptly competent in mathematics often fail to pursue mathematics-related careers because they have low self-efficacy perceptions about their competence" (Zeldin & Pajares, 2000).

Self-efficacy, often referred to as confidence in one's ability to succeed at a specific task, is a significant predictor of both the level of motivation for a task and ultimately task performance (Bandura & Locke, 2003). Self-efficacy influences interests, goals, performance and persistence (Eccles, 1994; Lent et al., 1994), and is associated with greater self-regulation, including more efficient use of problem-solving strategies and management of working time (Zimmerman, 2000). In addition to expending greater effort, efficacious individuals set more challenging goals, and persist longer to complete a task, even when faced with obstacles (Pajares, 2005; Zimmerman, 2000). This underlying belief in one's abilities impacts the choices students make concerning their field of study in tertiary education, which then determines their career choice (Zeldin et al., 2008).

Several studies have shown that male students hold higher self-efficacy than female students in maths and science. Reilly et al. (2019) found that, in general, girls have lower science self-efficacy beliefs across nations despite there being no significant difference in global science achievement scores. Similarly, a study among 121 grade 11 high school students showed that girls rated themselves as significantly less competent than boys in both mathematics and science (Hand et al., 2017). Perez-Felkner et al. (2017) found that grade 10 boys hold a growth mindset more often than girls and perceive their mathematics ability to be stronger than girls do, even when controlling for observed mathematics ability and other key predictors. They state that these gendered patterns hold even among the most mathematically talented students (boys and girls together), supporting the argument that ability beliefs and their influence cannot be explained by differences in innate talent. However, in contrast, Louis and Mistele (2012) found the science self-efficacy levels between 7,377 eighth-grade male and female students were not statistically different. Furthermore, Pajares et al. (2000) predicted higher science achievement scores for girls than for boys as girls had higher self-efficacy ratings. Such findings suggest lower female self-efficacy in science is not inevitable.

Self-image/Concept

Several studies have shown that academic self-concept is one of the most relevant determinants in students' selection of secondary school majors (Nagy et al., 2008). Similar mechanisms seem to be crucial for choosing study paths in higher education and for career choices (Nagy et al., 2006). The perceived closeness (between the self and a school subject) has been found to be predictive of students' career choice intentions (Hannover & Kessels, 2004; Kessels et al., 2006), and indeed, in relation to decisions to engage with science, data indicates that these are highly dependent on the 'image' students have of science and scientists (OECD, 2007).

Students' mathematics and science self-concepts have also been found to positively influence mathematics and science achievement (Jansen et al., 2014; Sahranavard et al., 2012; Wang et al., 2008). Unfortunately, some studies have found lower female than male self-concepts, often fuelled by gender stereotypes, even among students with high achievement in STEM subjects (see, for example, Ertl et al., 2017). In a study involving participants of a German Junior Science Olympiad, female students reported lower self-concepts than those of their male counterparts, despite receiving tangible feedback of being equally able and successful (Höffler et al., 2017). Perhaps not surprisingly, for male students, gender-science stereotyping seems to have a positive effect on self-concept and, thus, boosts their career aspirations in STEM (Makarova et al., 2019).

Attitudes and Enjoyment

Attitudes also seem to play an important role and can predict later academic performance and course-enrolment decisions (Barnes et al., 2005; Leibham et al., 2013). Some researchers have argued that these are more important than aptitude for explaining the underrepresentation of women in STEM (Else-Quest et al., 2013; Smeding, 2012), implying women have fewer positive attitudes to these fields than men.

Boys have been found to report more positive attitudes towards learning science than girls (Else-Quest et al., 2010; Reilly et al., 2019; Weinburgh, 1995). However, a wide variation between countries in boys' and girls' attitudes towards science has also been found, suggesting these attitudes are highly culturally bound (Reilly et al., 2019). Fifteen-year-old male students from 70 countries also reported greater interest in learning science than females, even though a majority of all students reported interest and enjoyment in the subject (OECD, 2016). Interest in science plays a pivotal role in motivating students to engage in science-related activities, to enrol in advanced science studies, and to work in a science-related field (Kang et al., 2019).

Other Barriers

Several other barriers have been discussed in relation to girls' achievement and participation in the sciences. These include the impact of teachers and role models, sociocultural factors and the ways in which science is assessed.

Teachers and Role Models

Many studies indicate that teachers, mentors and other adult role models are highly influential in shaping students' self-efficacy, attitudes, achievement and persistence in STEM disciplines (Gunderson et al., 2012; Hand et al., 2017). Parents and teachers can influence a child's view of science through attitudes and the provision of positive science experiences (Bhanot & Jovanovic, 2009; Crowley et al., 2001; Frome & Eccles, 1998; Lynch, 2002). Conversely, negative science classroom experiences can reduce science career aspirations and the pursuit of science in higher education (Archer et al., 2020).

Teachers also play a significant role as a source of support for students during the school years, which is a crucial time of identity formation for students. Teachers have been found to expect boys and girls to perform differently in STEM disciplines based on the characteristics they associate with being successful in these fields, as discussed previously. This may affect a teacher's behaviour towards students, which could, in turn, influence students' identity development (Hand et al., 2017).

A teacher's own self-efficacy or interests may also influence how STEM subjects are perceived. Students' career aspirations in science are likely to start at secondary school or before, (Kang et al., 2019), therefore messages primary school students receive regarding STEM subjects likely play an important role. In general, science is an area in which primary school teachers indicate low self-efficacy (Cakiroglu et al., 2012; Kazempour & Sadler, 2015), particularly female teachers (Lumpe et al., 2012), who often make up the majority of primary school teachers. As such, if female teachers show a lack of confidence or lack of interest in any or all STEM-related subjects, their female students may consider the subjects to be less suitable for them to study which may, eventually, decrease the probability of female students' career aspirations in the sciences.

In a study involving over 1500 grade eight, grade 10, and university-level students, Fouad et al. (2010) found that teachers were a leading source of both support and barriers in science. Lack of inspiration by teachers and lack of advice from teachers were seen as barriers to science coursework or STEM-related career decisions by the female grade eight and university students, yet the perception that teachers wanted them to do well was viewed as a strong source of psychological support for both male and female grade eight and 10 students. Traditionally, public schools and universities in the GCC are segregated by gender, although this is starting to change. Single-gender classes may alleviate comparisons in teachers' treatment of boys and girls but will not prevent the self-efficacy and interests of, and level of support from teachers, from impacting students.

Sociocultural Factors

In addition to the media and other influencers mentioned previously, children's views of learning science are shaped by various sociocultural factors, as well as their own inherent interests and talents (Reilly et al., 2019). Role congruity theory (Eagly & Karau, 2002) suggests that when people act counter to the gender role prescribed for their own gender, there could be potential repercussions. Since society, teachers included, tends to associate STEM with masculine characteristics, a girl performing well in STEM, or having the desire to pursue STEM could be subject to negative consequences and differential treatment (Hand et al., 2017). As teacher support has been identified as a source of increased self-efficacy and support in STEM subjects for girls (Fouad et al., 2010; Rice et al., 2013), differential treatment stemming from gender role biases of teachers and students could discourage positive self-efficacy in these fields (Hand et al., 2017).

There is also considerable diversity in the rights and status of women cross-culturally, which can impact their degree of participation in educational and occupational choices. PISA results have shown an association between gender equity and the gender gap when gender differences in mathematics achievement are examined; smaller gender differences were observed in more gender-equal countries (Guiso et al., 2008; Reilly, 2012), supporting the gender segregation hypothesis. This suggests that pronounced gender inequality may negatively influence the cognitive and educational development of girls (Baker & Jones, 1993).

Assessment of the Sciences

It may also be that assessment strategies in the sciences have fuelled the perception that 'science is for boys', by favouring boys in format and item construction. Paying attention to gender differences in performance when screening items in science test construction processes has been reported as a way to reduce, to some extent, the magnitude of observed score differences between the genders (Beller & Gafni, 1996). Dimitrov (1999) also pointed out that previous results about gender differences in science achievement, which showed males outperforming females, could be due to traditional item response formats on tests. He found that for high-ability students, boys outperformed girls on open-ended format items for physical science. This contrasted with previous findings that boys outperformed girls on multi-choice items, whereas girls did as well or better than boys on open-ended items (DeMars, 1998), but seems to support the notion that item format impacts performance differentially across genders. More recently, small relationships between science test item features and performance by gender were found in a study with almost 1500 undergraduate students (Federer et al., 2016), and in a study involving mathematics and English tests, test item format explained approximately 25% of the variation in gender achievement gaps (Reardon et al., 2018). Additionally, test anxiety has been repeatedly found to negatively correlate to performance (see Hembree, 1988; Ironsi, 2020; Schillinger et al., 2021), and was cited as a barrier in science by high school females in a large-scale study in the U.S. (Fouad et al., 2010).

Current State of the Gap

To ascertain the current state of the gender gap in STEM, specifically in the GCC, a range of student outcomes must be considered, and how they differ between boys and girls, if at all. The impact of recent outcomes on STEM enrolment in higher education is also discussed.

Student Outcomes

Student achievement outcomes in science for boys and girls appear to be much closer than they were a couple of decades ago but remain mixed. The lack of consistency in findings may partially reflect differences in the size and type of samples that are recruited, and what, specifically, was being measured. The most reliable evidence for gender differences in cognitive abilities is likely to come from samples that are representative of the general population (Hedges & Nowell, 1995), such as large-scale educational assessments of students on standardised tests. For example, if we consider the results of different Trends in International Mathematics and Science Study (TIMSS) iterations, an educational research study on student achievement in mathematics and science around the world involving students from now over 60 countries, we can see trends in the gender performance gap.

In 2007, TIMSS data for a representative sample of over 7000 students in the U.S. showed a statistically significant difference in the science achievement scores between female and male students, where the males' achievement scores were higher (Louis & Mistele, 2012). By 2011, small- to medium-sized gender differences were found for most individual nations in grade 8 mathematics and science achievement, although the direction varied and there were no global gender differences overall

(Reilly et al., 2019). Boys still reported more favourable attitudes and higher self-efficacy beliefs towards mathematics and science than girls (Reilly et al., 2019). In the most recent TIMSS data for grade 8 students, collected in 2019, girls had higher average science achievement in more countries than the reverse, and there were considerable levels of gender equity in average mathematics achievement (Mullis et al., 2020).

When we look at the GCC region specifically, girls achieved statistically significantly higher TIMSS 2019 scores than boys in all six countries for science. Oman showed the biggest gender performance gap (favoured towards girls), when ranked by the difference in average science achievement, of all 39 participating countries; with Bahrain and Saudi Arabia third and fourth respectively (Mullis et al., 2020). Kuwait, Qatar, and the UAE ranked fifth, sixth, and eighth respectively. Similarly, GCC girls achieved higher scores than boys for maths in all six countries, statistically significantly so in three countries: Oman, Bahrain, and Saudi Arabia. Students completing the TIMSS were also asked to rate how much they liked science, how confident they felt in science, and how much they valued it. The below tables show the results for Grade 8 boys and girls in the GCC (Table 1.1).

Overall, these results are positive, with more than 80% of all students 'very much' or 'somewhat' liking science. In all GCC countries, girls reported liking science more than boys did (Table 1.2).

In contrast to past research, girls reported a higher rate of confidence than boys in GCC countries, with the exception of Kuwait. On average, nearly a third of boys reported feeling not confident in science. Almost a quarter of girls reporting a lack of confidence is also of concern, given what has been discussed earlier regarding the positive relationship between confidence and pursuing STEM-related courses and careers (Table 1.3).

Table 1.1 Students like learning science -TIMSS, 2019, (Mullis et al., 2020)

Country (position)	Very much like learning science % Students		Somewhat like learning science % Students		Do not like learning science % Students	
	Boys	Girls	Boys	Girls	Boys	Girls
Bahrain (8)	40.12	48.40	40.13	39.69	19.75	11.91
Kuwait (18)	40.52	50.71	41.89	36.19	17.59	13.09
Oman (9)	38.58	52.96	49.92	38.67	11.50	8.37
Qatar (20)	36.53	40.04	46.15	43.12	17.33	16.82
Saudi Arabia (13)	41.49	49.98	44.40	36.60	14.11	13.42
United Arab Emirates (11)	42.17	46.92	42.64	38.83	15.20	14.24
GCC mean	39.90	48.17	44.19	38.85	15.91	12.98

Source Mullis et al. (2020)
Note (#) = position when all 39 participating countries are ranked by their average score on the Students Like Learning Science scale

Table 1.2 Students' confidence in science - TIMSS, 2019, (Mullis et al., 2020)

General/Integrated Science Country (position)	Very confident in science % Students		Somewhat confident in science % Students		Not confident in science % Students	
	Boys	Girls	Boys	Girls	Boys	Girls
Bahrain (4)	29.73	42.00	36.68	39.03	33.59	18.97
Kuwait (8)	34.28	26.45	42.57	41.86	23.16	31.69
Oman (9)	23.03	34.89	44.80	46.28	32.17	18.83
Qatar (11)	25.23	31.34	37.52	40.06	37.24	28.60
Saudi Arabia (6)	27.41	38.88	43.40	40.16	29.19	20.96
United Arab Emirates (10)	27.53	30.34	39.24	42.21	33.23	27.45
GCC mean	26.56	35.29	40.58	41.72	32.85	22.99

Source Mullis et al. (2020)
Note (#) = position when all 39 participating countries are ranked by their average score on the Students Confidence in Science scale

Table 1.3 Value placed upon learning science by students - TIMSS, 2019, (Mullis et al., 2020)

Country (position)	Strongly value science % Students		Somewhat value science % Students		Do not value science % Students	
	Boys	Girls	Boys	Girls	Boys	Girls
Bahrain (9)	48.24	54.24	36.56	34.70	15.21	11.01
Kuwait (7)	52.60	55.67	34.60	32.72	12.80	11.61
Oman (3)	52.95	61.47	37.90	31.97	9.15	6.56
Qatar (11)	47.89	49.43	37.60	36.05	14.51	14.52
Saudi Arabia (6)	53.38	55.97	34.81	33.11	11.81	10.92
United Arab Emirates (5)	53.74	55.82	35.18	33.52	11.08	10.65
GCC mean	51.47	55.44	36.11	33.68	12.43	10.88

Source Mullis et al. (2020)
Note (#) = position when all 39 participating countries are ranked by average score on the Students Value Science scale

Pleasingly, among all GCC countries, Grade 8 students largely reported valuing science. Again, the girls rated the value of science more highly than boys, with the exception of Qatar, where ratings were relatively even between the genders. These results are pleasing, as the extent to which students believe a field is valuable, important, and useful, influences which courses they choose and careers they pursue (Eccles, 1994; Reilly et al., 2019; Riegle-Crumb et al., 2011).

College and Careers

Despite the lack of a gender gap in achievement in science, the number of females studying STEM courses post-secondary and pursuing STEM careers remain significantly below those of males. With the 2011 TIMSS data showing no gender differences in achievement overall, we should be seeing this reflected in even gendered STEM university enrolments by now. So, while it is positive to find that GCC students, in general, reported liking, valuing and feeling confident in science, these findings are not indicative of a desire to pursue a career in the sciences.

The Global Gender Gap Report of the World Economic Forum (WEF) states that women are still underrepresented in the STEM fields (WEF, 2017, p. 31). Moreover, the occupational aspirations of 15-year-old adolescents suggest that gender segregation in the education and labour market will remain persistent (OECD, 2017). Positively, the male-to-female ratio among U.S. college majors in biology, chemistry, and mathematics is now about 1-to-1, however, in physics, engineering, and computer science, the ratio appears to have plateaued at about 4-to-1 (Cheryan et al., 2017; National Science Board, 2018). Gender gaps in these fields mean the fields themselves are missing out on the benefits of having gender diversity, including greater innovation, creativity, and collective intelligence (Page, 2007; Woolley et al., 2010) and women may be missing out on careers that are lucrative and high in status (Graves, 2014). These gaps can no longer be attributed to gender-related differences in ability.

Stereotypical beliefs about mathematics and science and low confidence still seem to be preventing young women from entering a STEM career (Lane et al., 2012; Makarova et al., 2019; Ramsey, 2017). Women who aspired to non-STEM majors as a field of study at university perceived mathematics, physics, and chemistry as more masculine than those who did choose STEM majors (Makarova et al., 2019). Women may similarly choose STEM courses less often than men because they have lower confidence in their mathematics abilities (Sax et al., 2015), or because they perceive a mismatch between their own self-image and the image of an occupation in the sciences (Bubany & Hansen, 2011; Gottfredson, 2002, 2005; Ratschinski, 2009).

Obviously, women may just have different career aspirations, and an important tenet of gender equity discourse needs to account for choice and agency. While it is important to make studying and working in STEM appealing and accessible to women, undue pressure to pursue STEM may be as counterproductive as the many barriers currently keeping women from the field, previously mentioned. However, recent research has suggested that women may simply have more options when it comes to choosing fields of study and potential careers. Breda and Napp (2019) claim women may have other options owing to a comparative advantage over men in English and/or reading, and Wang et al. (2013) found evidence of a greater likelihood that females with high mathematics ability also have high verbal ability and thus can consider a wider range of occupations than their male peers with high mathematics ability, who are more likely to have moderate verbal ability. What is not yet clear, and therefore worthy of further investigation, is why women with high abilities in both mathematical and verbal skills do not pursue STEM studies and careers.

Positively, in a recent study in the UAE with 66 grade five and six students, 61% indicated that they would like to either be a scientist or have a career that involves science in some way (Dickson et al., 2021). Previously unpublished data from the study showed that while the percentage of boys with science career aspirations was slightly higher than those of girls (67% vs 61%), the results reflected an even proportion of the boys and girls who reported enjoying the subject. When asked to elaborate on what they might like to do, the range of science involved in the desired occupations was broad. This was an encouraging finding, as previous literature has suggested that other than medicine, most children know very little about the careers and lives of those working in STEM disciplines, and this can act as a barrier, particularly to young people from traditionally underrepresented backgrounds (Perez-Felkner et al., 2017). Interestingly, the career aspirations did not seem to follow gender lines, with both boys and girls reporting interests in, among other things, medicine, technology, science education and working with animals.

Discussion and Recommendations

'Science is for boys' may still be a persistent belief in the GCC and beyond and, given that this has been held to be true for so long, it might take some time to change. This is true despite research showing that gender gaps in achievement are trivial, and gaps in other student outcomes are also closing, or even favouring girls. Assessments such as the TIMSS, referenced earlier, cannot tell the whole story, as we must consider the innumerable variables that play a part in how well a student might perform on such an assessment. Exposure to science, the way it is taught and assessed, self-efficacy, and stereotype endorsement or threat can all impact the way a student feels about, and performs in, the subject and these variables may act differently upon boys and girls. Nevertheless, the information from the latest TIMSS and the science aspirations of a small group of students in Abu Dhabi are promising.

Previous literature points to some areas to be mindful of in relation to stubborn stereotypes around science, boosting the self-efficacy of girls, the influence of teachers and how girls can be best supported. These areas are discussed here.

Overcoming Stereotypes

A critical evaluation of the image of science in schools might be one way to overcome gender-image-driven barriers to career aspirations of both female and male students. Gender stereotypical images of mathematics and science subjects are most likely internalised well before students need to make subject and career choices, so reflections on these should be encouraged from early childhood (Kang et al.,

2019; Makarova et al., 2019). Since stereotypical representations in textbooks influence secondary school students' understanding of, and feelings about, science (Good et al., 2010), an effort needs to be made to portray diverse representations in books at all educational levels. Gender stereotypical beliefs held by teachers, parents and other influencers who are involved in the development of vocational interests among children and secondary students must also be challenged (Makarova et al., 2019). Given that the students who participated in a recent UAE study (Dickson et al., 2021) had a comparable ratio of girls and boys who enjoyed science to those who wanted to pursue a science-related career, the teachers and parents of the participants may be doing a good job in this regard. A study into the perceptions of science and scientists among parents and teachers in the same community is recommended. Whether these adults are the reason behind healthy percentages of their children/students interested in science-related careers, or whether the students are interested despite these adults, such a study should provide interesting recommendations.

Makarova et al. (2019) found that young women who chose STEM as a field of study at university perceived mathematics, physics and chemistry as less masculine than young women who chose other majors. They also found that young men that did not fit the masculine stereotype also opted for non-STEM majors. This suggests it is important for both genders to be able to associate their self-image with that of an academic subject, for them to consider pursuing that subject. Addressing misperceptions around STEM careers, and emphasising potential benefits, may also help. In order to make science subjects appealing to many students, it is important to highlight the impact of science on their life, the fact that STEM careers hold the key to helping people and society, and how scientists balance their work and life without giving up their personal time (Kang et al., 2019).

Boosting Self-efficacy

Educators should be cognisant of the impact that STEM education can have on self-efficacy and intentionally cultivate an environment that promotes self-efficacy (Roche & Manzi, 2019). Indeed, several studies have shown that higher self-efficacy has been positively correlated with the uptake of STEM studies (see, for example, Nix et al., 2015; Perez-Felkner et al., 2017). Girls have been repeatedly found to have lower science self-efficacy beliefs than boys even when there is no significant difference in science achievement (see for example, Hand et al., 2017; Perez-Felkner et al., 2017; Reilly et al., 2019). Therefore, designing and implementing interventions to increase their self-efficacy should increase the number of girls who will consider studying and pursuing careers in STEM. If just a small percentage of these girls choose STEM over other studies and careers, this will help to reduce the gender gap. Fortunately, the literature points to some possible avenues for improving girls' self-efficacy for STEM. Decades of research have found girls underrate their abilities on tasks and careers that are culturally considered male (Perez-Felkner, 2018), therefore abolishing gender stereotypes should have a positive impact on girls' confidence in traditionally masculine subjects. Another study that found girls' have more negative perceptions of their ability under challenge, suggesting difficulty with learning that may lead girls away from scientific majors and careers (Perez-Felkner et al., 2017), implies that interventions to help build resilience may help. Feeling challenged is a normal and necessary part of learning, but students appear to experience optimal learning when their skill, interest and level of challenge are balanced (Schneider et al., 2016). Therefore, teachers should strive to find this balance.

Cultivating a growth mindset (Dweck, 2006) may also prove useful in the development of positive self-efficacy. A growth mindset acknowledges that mistakes are not failures but are an important part of the learning process. When studying the impact of a girls' STEM camp, Hughes and Roberts (2019) found that being open to challenge played a crucial role in how students felt about themselves in relation to STEM learning. They suggest that informal science education spaces and experiences can give space and time for learners to make mistakes and work through authentic science opportunities. Teachers in formal education can do the same, by providing a safe learning environment in which learners are encouraged to try new things.

Interventions that have resulted in positive results for girls' STEM self-efficacy to date include enrichment programmes (Ogle et al., 2017), robotics and game design (Leonard et al., 2016), and a game design workshop (Cakir et al., 2017). Portnoy and Schrier (2019) implemented a deck-building card game with 178 middle school students and found that a higher proportion of females than males reported an interest in being a scientist after the game, and those whose scores changed more substantially from pre- to post- also demonstrated a higher interest in being a scientist. They suggest that the game places students in the role of scientist and in doing so, improves their efficacy and their belief that they can take on the role of a scientist. The Science Research & Engineering Program, established in 1998 to engage and expose the all-female students at a secondary school, and tracked in a longitudinal study across 20 years, found that students who participated in the programme were more likely to pursue a major in a STEM field and continue on to a STEM occupation compared to students who did not participate, national female averages, and national subsets. Participants attributed their outcomes, in part, to an increase in confidence (Hunt et al., 2021).

These successful interventions and the other possible avenues for increasing the STEM self-efficacy of girls discussed above provide for a hopeful future for females in STEM. The beginning of secondary school has been highlighted as a critical time for STEM selection (Hunt et al., 2021; Sadler et al., 2012), therefore a positive STEM experience and healthy self-efficacy during this important period could potentially guide students in pursuit of STEM majors and occupations and prevent girls who may be interested in STEM from shifting away from STEM in college or beyond (Hunt et al., 2021).

The Influence of Teachers

Teacher support has been found to be highly important in encouraging students to pursue science and other STEM subjects. Teachers can play a pivotal role in providing encouragement for girls pursuing STEM, especially where female role models are wanting (Hand et al., 2017), and teacher support in mathematics and science has been correlated with increased student self-efficacy and positive perceptions of these subjects (Rice et al., 2013). Teachers have also been found to influence a child's view of science through attitudes, support and science-positive experiences (see, for example, Frome & Eccles, 1998; Gunderson et al., 2012; Hand et al., 2017; Hunt et al., 2021; Vygotsky, 1980). Findings from multiple studies suggest a clear alignment between science learning experiences and interest in STEM subjects (see, for example, Anderhag et al., 2014; Hasni & Potvin, 2015; Kang & Keinonen, 2018; Logan & Skamp, 2013). Making connections between content and 'real-world' science issues, using guided inquiry-based learning and implementing practical activities are teacher activities that have been linked with positive student outcomes, such as performance, interest and enjoyment (Jiang & McComas, 2015; Kang & Keinonen, 2018; Logan & Skamp, 2013; Minner et al., 2010). Teachers should be careful to choose curriculum materials, such as textbooks and videos, that indicate less gender-biased images or texts (Logan & Skamp, 2013). In addition to a teacher's pedagogy, their characteristics can also influence how students perceive science and impact their interest in the subject. In year seven to 10 science classes, good relationships with students, enthusiasm, use of humour, and fun, relaxed lessons positively impacted students, while sarcasm and a stressed learning environment caused disaffection (Logan & Skamp, 2013). These things must be considered seriously; teacher influence was found to be a statistically significant predictor for grade six to 12 girls' interest and confidence in both mathematics and science in a study involving 1283 female students (Heaverlo et al., 2013). Other classroom occurrences also need to be considered and addressed. For example, in hundreds of classroom observations, male students regularly monopolised classroom conversations, asked and answered more questions, received more praise, and received more help (Sadker et al., 2009). While these microinequities are usually unintentional, teachers must be cognisant of any gender bias in their actions and perceive the extent to which their modelling can influence students' future interest in science and even future career choices (Logan & Skamp, 2013; Sadker et al., 2009). It is recommended that teachers practice self-reflection to discover and mediate any biases regarding gender to minimise negative impacts on student achievement and interest. It is also recommended that pre-service teachers in the GCC are provided opportunities to learn about gender inequalities in STEM during their teacher education. Given that assessment formats have been found to impact performance differentially across genders (see, for example, Federer et al., 2016; Reardon et al., 2018), it is further recommended that teachers ensure that varied assessments are implemented in STEM-related classes so that all students can be evaluated fairly.

Role Models and Other Supports

Women role models in STEM for girls interested in pursuing these fields are important (Bettinger & Long, 2005; Moorhouse, 2017), and increasing visibility and access to women scientists, both fictional and real, is suggested as an approach to supporting girls' interest and engagement through middle and high school (Perez-Felkner et al., 2017). Exposing 12–16-year-old girls to the professional and personal experiences of actual female role models with a successful professional trajectory in STEM fields significantly increased enjoyment of, and the perceived importance of, these fields (Gonzalez-Perez, 2020). Furthermore, the role model sessions increased girls' expectations of success in mathematics and decreased the effect of gender-role stereotypes. "In the imagined worlds created by entertainment media, STEM portrayals can immediately be gender equitable" (The Lyda Hill Foundation & Geena Davis Institute on Gender in Media, 2018, p. 12). The media often reflects current societal beliefs and attitudes but can also influence them. It is recommended that stakeholders interested in encouraging females into STEM apply pressure on media outlets to use its considerable power by presenting equity and challenging existing gender biases in STEM. Although this chapter does not examine cultural influences on the uptake of STEM, it is presumed that seeing Arab female role models would be additionally beneficial for female students in the GCC.

Finding ways to help girls envisage themselves in STEM careers may also support more girls to pursue such a pathway. For example, Kang et al. (2019) found that grade seven girls considered money, fame or a high position to be important for career satisfaction. Thus, they recommended introducing STEM careers in and out of school environments in which students could be exposed to scientists who have achieved such outcomes, so that girls may see science as being more relevant for their future and therefore be more interested in it (Kang et al., 2019). Research into the specific elements considered important for career satisfaction for GCC girls is recommended.

Social network research on course-taking patterns indicated that girls are particularly likely to follow their same-gender friends into or out of STEM preparatory coursework in high school (Riegle-Crumb et al., 2006), suggesting that girls, themselves, can be influencing role models. Challenging gender stereotypes, and having supportive teachers and positive role models, may therefore, only need to encourage a few girls into the sciences; they will then draw their friends with them. Students of both genders benefit from having peer support in STEM; however, peer support may be more important for females than males (Ost, 2010; Riegle-Crumb et al., 2006). In a large study ($n = 5025$) over three years starting when students were 14–15 years old, both friends and other peers in the classroom influenced preferences towards STEM subjects (Raabe et al., 2019). The authors found strong evidence that students adjusted their preferences to those of their friends, and that the effect of girls' STEM preferences on other girls' STEM preferences is highly significant and larger than on general subject preferences (Raabe et al., 2019). Conversely, Fouad et al. (2010) found that having friends that were not interested in science was a notable barrier for girls (and boys) at the grade 8 level. The significance of peer support appears to continue into higher education. When women in STEM can form supportive connections with peers in their area of study, they tend to experience increased confidence, motivation and achievement (Dasgupta et al., 2015; Robnett & Leaper, 2013; Robnett & Thoman, 2017; Walton et al., 2012). Robnett and Thoman (2017) found that a group of female undergraduates that they labelled "self-doubting achievers" (characterised by low success expectancies despite their relatively strong academic achievement) consistently reported lower levels of support from their STEM peers and were therefore at a greater risk of exiting the STEM pipeline. They suggested that interventions that foster a sense of social belongingness may be helpful for these women, but it is likely such initiatives would be useful for all females in STEM, starting well before they reach higher education. Social belongingness can be instilled through exposure to female peers who are excelling in STEM (Dasgupta et al., 2015; Stout et al., 2011), or by establishing peer mentoring programmes (Robnett & Thoman, 2017).

Finally, qualitative, longitudinal research that tracks GCC students' science aspirations across high school, into tertiary education and beyond, is recommended. This would help to explain why students choose, or don't, to pursue a career in the sciences, the impact of support on those choices, and where any persistent barriers may still exist.

Conclusion

We cannot entice all students into STEM fields, nor would this be a desirable or just outcome. However, encouraging more women into studies and careers where they are underrepresented remains an important undertaking. Research shows diversity increases the quality of scientific work, helping generate more innovative and influential ideas (Charlesworth & Banaji, 2019; Freeman & Huang, 2014; Page, 2007; Woolley et al., 2010). Furthermore, the more women that enter STEM fields, the more role models there are for younger women, potentially creating an increasing cycle.

In the GCC region, girls out-achieved boys on the most recent TIMSS iteration (Mullis et al., 2020). GCC girls also reported liking and valuing science more than boys and reported a higher rate of confidence than boys (Mullis et al., 2020). Over two-thirds of 10–11-year-old girls interviewed in Abu Dhabi expressed an interest in a variety of science-related careers (Dickson et al., 2021). These students were the same age as the next cohort who will complete the TIMSS, therefore allowing for some tentative predictions to be made for the next iteration, and an optimistic outlook for female participation in STEM in the GCC moving forward.

References

Anderhag, P., Hamza, K. M., & Wick, P. (2014). What can a teacher do to support students' interest in science? A study of the constitution of taste in a science classroom. *Research in Science Education, 45*, 749–784. https://doi.org/10.1007/s11165-014-9448-4.

Archer, L., DeWitt, J., Osborne, J., Dillon, J., Willis, B., & Wong, B. (2010). "Doing" science versus "being" a scientist: examining 10/11-year-old school children's constructions of science through the lens of identity. *Science Education, 94*, 617–639. https://doi.org/10.1002/sce.20399

Archer, J., & MacRae, M. (1991). Gender perceptions of school subjects among 10–11-year-olds. *British Journal of Educational Psychology, 61*, 99–103. https://doi.org/10.1111/j.2044-8279.1991.tb00965.x

Archer, L., Moote, J., Macleod, E., Francis, B., & DeWitt, J. (2020). *ASPIRES 2: Young people's science and career aspirations, age 10–19.* UCL Institute of Education: London, UK.

Baker, D. P., & Jones, D. P. (1993). Creating gender equality: Cross-national gender stratification and mathematical performance. *Sociology of Education, 66*(2), 91–103. https://doi.org/10.2307/2112795

Bandura, A., & Locke, E. A. (2003). Negative self-efficacy and goal effects revisited. *Journal of Applied Psychology, 88*, 87–99. https://doi.org/10.1037/0021-9010.88.1.87

Barnes, G., McInerney, D. M., & Marsh, H. W. (2005). Exploring sex differences in science enrolment intentions: An application of the general model of academic choice. *The Australian Educational Researcher, 32*(2), 1–23. https://doi.org/10.1007/BF03216817

Bell, K. N., & Norwood, K. (2007). Gender equity intersects with mathematics and technology: Problem- solving education for changing times. In D. Sadker & E. S. Silber (Eds.), *Gender in the classroom* (pp. 225–258). Lawrence Erlbaum Associates.

Beller, M., & Gafni, N. (1996). The 1991 international assessment of educational progress in mathematics and sciences: the gender differences perspective. *Journal of Educational Psychology, 88*(2), 365–377.

Bettinger, E. P., & Long, B. T. (2005). Do faculty serve as role models? The impact of instructor gender on female students. *Understanding Teacher Quality, 95*(2), 152–157.

Beyer, S. (1990). Gender differences in the accuracy of self-evaluations of performance. *Journal of Personality and Social Psychology, 59*(5), 960–970. https://doi.org/10.1037/0022-3514.59.5.960

Bhanot, R. T., & Jovanovic, J. (2009). The links between parent behaviors and boys' and girls' science achievement beliefs. *Applied Developmental Science, 13*(1), 42–59. https://doi.org/10.1080/10888690802606784

Breda, T., & Napp, C. (2019). Girls' comparative advantage in reading can largely explain the gender gap in math-related fields. *Proceedings of the National Academy of Sciences, 116*(31), 15435–15440.

Brown, E. R., Thoman, D. B., Smith, J. L., & Diekman, A. B. (2015). Closing the communal gap: The importance of communal affordances in science career motivation. *Journal of Applied Social Psychology, 45*, 662–673.

Bubany, S. T., & Hansen, J. I. C. (2011). Birth cohort change in the vocational interests of female and male college students. *Journal of Vocational Behavior, 78*, 59–67. https://doi.org/10.1016/j.jvb.2010.08.002

Cakir, N. A., Gass, A., Foster, A., & Lee, F. J. (2017). Development of a game design workshop to promote young girls' interest towards computing through identity exploration. *Computers & Education, 108*, 115–130.

Cakiroglu, J., Capa-Aydin, Y., & Hoy, A. W. (2012). Science teaching efficacy beliefs. In *Second international handbook of science education* (pp. 449–461). Springer, Dordrecht.

Carli, L. L., Alawa, L., Lee, Y., Zhao, B., & Kim, E. (2016). Stereotypes about gender and science: Women ≠ scientists. *Psychology of Women Quarterly, 40*(2), 244–260. https://doi.org/10.1177/0361684315622645

Chambers, D. W. (1983). Stereotypic images of the scientist: The draw-a-scientist test. *Science Education, 67*(2), 255–265.

Charlesworth, T. E. S., & Banaji, M. R. (2019). Gender in science, technology, engineering, and mathematics: Issues, causes solutions. *Journal of Neuroscience, 39*(37), 7228–7243. https://doi.org/10.1523/JNEUROSCI.0475-18.2019

Cheryan, S., Ziegler, S. A., Montoya, A. K., & Jiang, L. (2017). Why are some STEM fields more gender balanced than others? *Psychological Bulletin, 143*(1), 1–35.

Cleary, A. (1991). Gender differences in aptitude and achievement test scores. In *Sex equity in educational opportunity, achievement, and testing: Proceedings of the 1991 ETS Invitational Conference* (pp. 51–90). Princeton, NJ: Educational Testing Service.

Crowley, K., Callanan, M. A., Tenenbaum, H. R., & Allen, E. (2001). Parents explain more often to boys than to girls during shared scientific thinking. *Psychological Science, 12*(3), 258–261. https://doi.org/10.1111/1467-9280.00347

Cvencek, D., Kapur, M., & Meltzoff, A. N. (2015). Math achievement, stereotypes, and math self-concepts among elementary-school students in Singapore. *Learning and Instruction, 39*, 1–10. https://doi.org/10.1016/j.learninstruc.2015.04.002

Dasgupta, N., Scircle, M. M., & Hunsinger, M. (2015). Female peers in small work groups enhance women's motivation, verbal participation, and career aspirations in engineering. *Proceedings from the National Academy of Sciences, 112*, 4988–4993. https://doi.org/10.1073/pnas.1422822112

Deaux, K., & LaFrance, M. (1998). Gender. In D. T. Gilbert, S. T. Fiske, & G. Lindzey (Eds.), *The handbook of social psychology* (pp. 788–827). McGraw-Hill.

DeMars, C. F. (1998). Gender differences in mathematics and science on a high school proficiency exam: The role of response format. *Applied Measurement in Education, 11*, 279–299.

Dickson, M., McMinn, M., Cairns, D., & Osei-Tutu, S. (2021). Children's perceptions of scientists, and of themselves as scientists. *International Journal on Math, Science and Technology Education, 9*, 643–669. https://doi.org/10.31129/LUMAT.9.1.1605.

Diekman, A. B., Brown, E. R., Johnston, A. M., & Clark, E. K. (2010). Seeking congruity between goals and roles: A new look at why women opt out of science, technology, engineering, and mathematics careers. *Psychological Science, 21*(8), 1051–1057.

Dimitrov, D. M. (1999). Gender differences in science achievement: Differential effect of ability, response format, and strands of learning outcomes. *School Science and Mathematics, 99*(8), 445–450.

Dweck, C. (2006). *Mindset: The new psychology of success*. Ballantine Press.

Eagly, A. H., & Karau, S. J. (2002). Role congruity theory of prejudice toward female leaders. *Psychological Review, 109*, 573–598.

Eccles, J. S. (1994). Understanding women's educational and occupational choices: Applying the Eccles et al. model of achievement-related choices. *Psychology of Women Quarterly, 18*, 585–609.

Else-Quest, N. M., Hyde, J. S., & Linn, M. C. (2010). Cross-national patterns of gender differences in mathematics: A meta-analysis. *Psychological Bulletin, 136*(1), 103–127. https://doi.org/10.1037/a0018053

Else-Quest, N. M., Mineo, C. C., & Higgins, A. (2013). Math and science attitudes and achievement at the intersection of gender and ethnicity. *Psychology of Women Quarterly, 37*(3), 293–309. https://doi.org/10.1177/0361684313480694

Engelhard, G. (1990). Gender differences in performance on mathematics items: Evidence from the United States and Thailand. *Contemporary Educational Psychology, 15*, 13–26.

Ertl, B., Luttenberger, S., & Paechter, M. (2017). The impact of gender stereotypes on the self-concept of female students in stem subjects with an underrepresentation of females. *Frontiers in Psychology, 8*(703). https://doi.org/10.3389/fpsyg.2017.00703.

Farrell, L., & McHugh, L. (2017). Examining gender-STEM bias among STEM and non-STEM students using the implicit relational assessment procedure (IRAP). *Journal of Contextual Behavioral Science, 6*(1), 80–90. https://doi.org/10.1016/j.jcbs.2017.02.001

Federer, M. R., Nehm, R. H., & Pearl, D. K. (2016). Examining gender differences in written assessment tasks in biology: A case study of evolutionary explanations. *CBE Life Sciences Education, 15*(1), ar2. https://doi.org/10.1187/cbe.14-01-0018.

Feingold, A. (1994). Gender differences in personality: A meta-analysis. *Psychological Bulletin, 116*(3), 429–456. https://doi.org/10.1037/0033-2909.116.3.429

Fennema, E. (2000). *Gender and mathematics: What is known and what do I wish was known?* Paper prepared for the Fifth Annual Forum of the National Institute for Science Education, Detroit, Michigan. Retrieved from http://archive.wceruw.org/nise/News_Activities/Forums/Fennemapaper.htm.

Fennema, E., Peterson, P. L., Carpenter, T. P., & Lubinski, C. A. (1990). Teachers' attributions and beliefs about girls, boys, and mathematics. *Educational Studies in Mathematics, 21*(1), 55–69. https://doi.org/10.1007/BF00311015

Fouad, N. A., Hackett, G., Smith, P. L., Kantamneni, N., Fitzpatrick, M., Haag, S., & Spencer, D. (2010). Barriers and supports for continuing in mathematics and science: Gender and educational level differences. *Journal of Vocational Behavior, 77*(3), 361–373. https://doi.org/10.1016/j.jvb.2010.06.004

Freeman, R. B., & Huang, W. (2014). Collaboration: Strength in diversity. *Nature, 513*, 305. https://doi.org/10.1038/513305a

Friedler, Y., & Tamir, P. (1990). Sex differences in science education in Israel: An analysis of 15 years of research. *Research in Science and Technological Education, 59*, 185–213.

Frome, P. M., & Eccles, J. S. (1998). Parents' influence on children's achievement-related perceptions. *Journal of Personality and Social Psychology, 74*(2), 435–452. https://doi.org/10.1037/0022-3514.74.2.435

Good, J. J., Woodzicka, J. A., & Wingfield, L. C. (2010). The effects of gender stereotypic and counter-stereotypic textbook images on science performance. *Journal of Social Psychology, 150*, 132–147. https://doi.org/10.1080/00224540903366552

González-Pérez, S., Mateos de Cabo, R., & Sáinz, M. (2020). Girls in STEM: Is it a female role-model thing? *Frontiers in Psychology, 11*(2204), 1–21. https://doi.org/10.3389/fpsyg.2020.02204

Gottfredson, L. S. (2002). Gottfredson's theory of circumscription, compromise, and self-creation. *Career Choice and Development, 4*, 85–148.

Gottfredson, L. S. (2005). Applying Gottfredson's theory of circumscription and compromise in career guidance and counselling. In S.D. Brown & R.W. (Eds.). *Career development and counselling: putting theory and research to work* (pp. 71–100). Wiley and Sons.

Graves, J. A. (2014, February 12). *The 25 best jobs of 2014: The top occupations to pursue for this year and this decade.* U.S. News. Retrieved from http://money.usnews.com/money/careers/slideshows/the-25-bestjobs-of-2014/13.

Guiso, L., Monte, F., Sapienza, P., & Zingales, L. (2008). Culture, gender, and math. *Science, 320*(5880), 1164–1165. https://doi.org/10.1126/science.1154094

Gunderson, E., Ramirez, G., Levine, S. C., & Beilock, S. (2012). The role of parents and teachers in the development of gender-related math attitudes. *Sex Roles, 66*(3), 153–166. https://doi.org/10.1007/s11199-011-9996-2

Halim, M. L., & Ruble, D. N. (2010). Gender identity and stereotyping in early and middle childhood. In J. C. Chrisler & D. R. McCreary (Eds.), *Handbook of gender research in psychology* (pp. 495–525). Springer.

Hand, S., Rice, L., & Greenlee, E. (2017). Exploring teachers' and students' gender role bias and students' confidence in STEM fields. *Social Psychology of Education, 20*, 929–945. https://doi.org/10.1007/s11218-017-9408-8

Hannover, B., & Kessels, U. (2004). Self-to-prototype matching as a strategy for making academic choices. Why high school students do not like math and science. *Learning and Instruction, 14*, 51–67. https://doi.org/10.1016/j.learninstruc.2003.10.002

Hasni, A., & Potvin, P. (2015). Student's interest in science and technology and its relationships with teaching methods, family context and self-efficacy. *International Journal of Environmental and Science Education, 10*(3), 337–366.

Heaverlo, C., Cooper, R., & Laanan, F. S. (2013). STEM development: Predictors for 6th-12th grade girls' interest and confidence in science and math. *Journal of Women and Minorities in Science and Engineering, 19*(2), 121–142. https://doi.org/10.1615/JWomenMinorScienEng.2013006464

Hedges, L. V., & Nowell, A. (1995). Sex differences in mental test scores, variability, and numbers of high-scoring individuals. *Science, 269*(5220), 41–45. https://doi.org/10.1126/science.7604277

Hembree, R. (1988). Correlates, causes, effects, and treatment of test anxiety. *Review of Educational Research, 58*(1), 47–77.

Höffler, T. N., Bonin, V., & Parchmann, I. (2017). Science vs. sports: Motivation and self-concepts of participants in different school competitions. *International Journal of Science and Mathematics Education*. Advance online publication. https://doi.org/10.1007/s10763-016-9717-y.

Hughes, R., & Roberts, K. (2019). The role of STEM self-efficacy on STEM identity for middle school girls after participation in a single-sex informal STEM education program. *International Journal of Gender, Science and Technology, 11*(2), 286–311.

Hunt, P. K., Dong, M., & Miller, C. M. (2021). A multi-year science research or engineering experience in high school gives women confidence to continue in the STEM pipeline or seek advancement in other fields: A 20-year longitudinal study. *PLoS ONE, 16*(11), e0258717. https://doi.org/10.1371/journal.pone.0258717

Hyde, J. S., Fennema, E., & Lamon, S. J. (1990). Gender differences in mathematics performance: A meta-analysis. *Psychological Bulletin, 107*(2), 139–155. https://doi.org/10.1037/0033-2909.107.2.139

Ironsi, C. S. (2020). Impact of test anxiety on test scores of preservice teachers in northern Cyprus. *International Journal of Educational Management and Development Studies, 1*(2), 19–36.

Jansen, M., Schroeders, U., & Lüdtke, O. (2014). Academic self-concept in science: Multidimensionality, relations to achievement measures, and gender differences. *Learning and Individual Differences, 30*, 11–21. https://doi.org/10.1016/j.lindif.2013.12.003

Jiang, F., & McComas, W. F. (2015). The effects of inquiry teaching on student science achievement and attitudes: Evidence from propensity score analysis of PISA data. *International Journal of Science Education, 37*, 554–576.

Jones, B., Ruff, C., & Paretti, M. (2013). The impact of engineering identification and stereotypes on undergraduate women's achievement and persistence in engineering. *Social Psychology of Education, 16*, 471–493. https://doi.org/10.1007/s11218-013-9222-x

Kang, J., Hense, J., Scheersoi, A., & Keinonen, T. (2019). Gender study on the relationships between science interest and future career perspectives. *International Journal of Science Education, 41*(1), 80–101. https://doi.org/10.1080/09500693.2018.1534021

Kang, J., & Keinonen, T. (2018). The effect of student-centered approaches on students' interest and achievement in science: Relevant topic-based, open and guided inquiry-based, and discussion-based approaches. *Research in Science Education, 48*, 865–885. https://doi.org/10.1007/s11165-016-9590-2

Kazempour, M., & Sadler, T. D. (2015). Pre-service teachers' science beliefs, attitudes, and self-efficacy: A multi-case study. *Teaching Education, 26*(3), 247–271.

Kerkhoven, A. H., Russo, P., Land-Zandstra, A. M., Saxena, A., & Rodenburg, F. J. (2016). Gender stereotypes in science education resources: A visual content analysis. *PLoS ONE, 11*(11), e0165037. https://doi.org/10.1371/journal.pone.0165037

Kessels, U. (2005). Fitting into the stereotype: How gender-stereotyped perceptions of prototypic peers relate to liking for school subjects. *European Journal of Psychology of Education, 20*, 309–323. https://doi.org/10.1007/BF03173559

Kessels, U., Rau, M., & Hannover, B. (2006). What goes well with physics? Measuring and altering the image of science. *British Journal of Educational Psychology, 76*, 761–780. https://doi.org/10.1348/000709905X59961

Konrad, A. M., Ritchie, J. E. J., Lieb, P., & Corrigall, E. (2000). Sex differences and similarities in job attribute preferences: A meta-analysis. *Psychological Bulletin, 126*, 593–641.

Lane, K. A., Goh, J. X., & Driver-Linn, E. (2012). Implicit science stereotypes mediate the relationship between gender and academic participation. *Sex Roles, 66*, 220–234. https://doi.org/10.1007/s11199-011-0036-z

Larose, S., Ratelle, C. F., Guay, F., Senécal, C., & Harvey, M. (2006). Trajectories of science self-efficacy beliefs during the college transition and academic and vocational adjustment in science and technology programs. *Educational Research and Evaluation, 12*, 373–393.

Leibham, M. B., Alexander, J. M., & Johnson, K. E. (2013). Science interests in preschool boys and girls: Relations to later self-concept and science achievement. *Science Education, 97*(4), 574–593. https://doi.org/10.1002/sce.21066

Lent, R. W., Brown, S. D., & Hackett, G. (1994). Toward a unifying social cognitive theory of career and academic interest, choice, and performance. *Journal of Vocational Behavior, 45*(1), 79–122. https://doi.org/10.1006/jvbe.1994.1027

Leonard, J., Buss, A., Gamboa, R., Mitchel, M., Fashola, O. S., Hubert, T., & Almughyirah, S. (2016). Using robotics and game design to enhance children's self-efficacy, STEM attitudes and computational thinking skills. *Journal of Science Education and Technology, 25*(6), 860–876.

Linn, M. C. (1985). Fostering equitable consequences from computer learning environments. *Sex Roles, 13*, 229–240.

Linn, M. C. (1991). Gender differences in educational achievement. In *Sex equity in educational opportunity, achievement, and testing: Proceedings of the 1991 ETS Invitational Conference* (pp. 11–50). Princeton, NJ: Educational Testing Service.

Logan, M. R., & Skamp, K. R. (2013). The impact of teachers and their science teaching on students' 'science interest': A four-year study. *International Journal of Science Education, 35*(17), 2879–2904. https://doi.org/10.1080/09500693.2012.667167

Long, M., Steinke, J., Applegate, B., Knight Lapinski, M., Johnson, M. J., & Ghosh, S. (2010). Portrayals of male and female scientists in television programs popular among middle school-age children. *Science Communication, 32*, 356–382. https://doi.org/10.1177/1075547009357779

Louis, R. A., & Mistele, J. M. (2012). The differences in scores and self-efficacy by student gender in mathematics and science. *International Journal of Science and Mathematics Education, 10*, 1163–1190 (2012). https://doi.org/10.1007/s10763-011-9325-9.

Loverock, B., & Hart, M. M. (2018). What a scientist looks like: Portraying gender in the scientific media. *FACETS, 3*(1), 754–763. https://doi.org/10.1139/facets-2017-0110

Lumpe, A., Czerniak, C., Haney, J., & Beltyukova, S. (2012). Beliefs about teaching science: The relationship between elementary teachers' participation in professional development and student achievement. *International Journal of Science Education, 34*(2), 153–166.

Lynch, J. (2002). Parents' self-efficacy beliefs, parents' gender, children's reader self-perceptions, reading achievement and gender. *Journal of Research in Reading, 25*(1), 54–67. https://doi.org/10.1111/1467-9817.00158

Maccoby, E. E., & Jacklin, C. N. (1974). *The psychology of sex differences*. Stanford University Press.

Makarova, E., Aeschlimann, B., & Herzog, W. (2019). The gender gap in STEM fields: The impact of the gender stereotype of math and science on secondary students' career aspirations. *Frontiers in Education, 4*(60). https://doi.org/10.3389/feduc.2019.00060.

Makarova, E., & Herzog, W. (2015). Trapped in the gender stereotype? The image of science among secondary school students and teachers. *Equality, Diversity and Inclusion: An International Journal, 34*, 106–123. https://doi.org/10.1108/EDI-11-2013-0097

Martinez, M. E., & Mead, N. A. (1988). *Computer competence*. Educational Testing Service.

Miller, D. I., Nolla, K. M., Eagly, A. H., & Uttal, D. H. (2018). The development of children's gender-science stereotypes: A meta-analysis of 5 decades of US draw-a-scientist studies. *Child Development, 89*, 1943–1955. https://doi.org/10.1111/cdev.13039

Minner, D. D., Levy, A. J., & Century, J. (2010). Inquiry-based science instruction—what is it and does it matter? Results from a research synthesis years 1984 to 2002. *Journal of Research in Science Teaching, 47*, 474–496. https://doi.org/10.1002/tea.20347

Moorhouse, E. A. (2017). Sex segregation by field of study and the influence of labor markets: Evidence from 39 countries. *International Journal of Comparative Sociology, 58*, 3–32.

Mullis, I. V. S., Martin, M. O. Foy, P., Kelly, D. L., & Fishbein, B. (2020). *TIMSS 2019 International results in mathematics and science*. TIMSS & PIRLS International Study Center, Boston College.

Nagy, G., Garrett, J., Trautwein, U., Cortina, K. S., Baumert, J., & Eccles, J. S. (2008). Gendered high school course selection as a precursor of gendered careers: the mediating role of self-concept and intrinsic value. In H. M. G. Watt and J. S. Eccles (Eds.), *Gender and Occupational Outcomes* (pp. 115–143). American Psychological Association. https://doi.org/10.1037/11706-004.

Nagy, G., Trautwein, U., Baumert, J., Köller, O., & Garrett, J. (2006).Gender and course selection in upper secondary education: effects of academic self-concept and intrinsic value. *Educational Research and Evaluation, 12*, 323–345. https://doi.org/10.1080/13803610600765687

National Assessment of Educational Progress. (1988). *The science report card: Elements of risk and recovery: Trends and achievement based on the 1986 National Assessment*. Educational Testing Service.

National Science Board. (2018). *Science & engineering indicators*. Retrieved from https://www.nsf.gov/statistics/2018/nsb20181/.

National Science Foundation. (2006). *Women, minorities, and persons with disabilities in science and engineering*. Retrieved from http://www.nsf.gov/statistics/wmpd/employ.cfm.

Nix, S., Perez-Felkner, L., & Thomas, K. (2015). Perceived mathematical ability under challenge: A longitudinal perspective on sex segregation among STEM degree fields. *Frontiers in Psychology*. https://doi.org/10.3389/fpsyg.2015.00530

Nosek, B. A., Banaji, M. R., & Greenwald, A. G. (2002). Math= male, me = female, therefore math ≠ me. *Journal of Personality and Social Psychology, 83*(1), 44–59. https://doi.org/10.1037//0022-3514.83.1.44

Nosek, B. A., Smyth, F. L., Sriram, N., Lindner, N. M., Devos, T., Ayala, A., Bar-Anan, Y., Bergh, R., Cai, H., Gonsalkorale, K., Kesebir, S., Maliszewski, N., Neto, F., Olli, E., Park, J., Schnabel, K., Shiomura, K., Tudor Tulbure, B., Wiers, R. W., … Greenwald, A. G. (2009). National differences in gender-science stereotypes predict national sex differences in science and math achievement. *Proceedings of the National Academy of Sciences of the United States of America, 106*, 10593–10597. https://doi.org/10.1073/pnas.0809921106.

OECD (2007). *PISA 2006: Science competencies for tomorrow's world. Volume 1: Analysis*. OECD.

OECD. (2016). *PISA 2015 results—Excellence and Equity in Education*. OECD Publishing.

OECD. (2017). *The pursuit of gender equality: An Uphill battle*. OECD Publishing. https://doi.org/10.1787/9789264281318-en

Ogle, J. P., Hyllegard, K. H., Rambo-Hernandez, K., & Park, J. (2017). Building middle school girls' self-efficacy, knowledge, and interest in math and science through the integration of fashion and STEM. *Journal of Family and Consumer Sciences, 109*(4), 33–40.

Ost, B. (2010). The role of peers and grades in determining major persistence in the sciences. *Economics of Education Review, 29*, 923–934. https://doi.org/10.1016/j.econedurev.2010.06.011

Page, S. E. (2007). *The difference: How the power of diversity creates better groups, firms, schools, and societies*. Princeton University Press.

Pajares, F. (2005). Gender differences in mathematics self-efficacy beliefs. In A. M. Gallagher & J. C. Kaufman (Eds.), *Gender differences in mathematics: An integrative psychological approach* (pp. 294–315). Cambridge University Press.

Pajares, F., Britner, S., & Valiante, G. (2000). Relation between achievement goals and self-beliefs of middle school students in writing and science. *Contemporary Educational Psychology, 25*(4), 406–422. https://doi.org/10.1006/ceps.1999.1027

Perez-Felkner, L. (2018, November 27). *The key to fixing the gender gap in math and science: Boost women's confidence*. Retrieved from https://phys.org/news/2018-11-key-gender-gap-mathscience.html.

Perez-Felkner, L., Nix, S., & Thomas, K. (2017). Gendered pathways: How mathematics ability beliefs shape secondary and postsecondary course and degree field choices. *Frontiers in Psychology, 8*(386), 1–11. https://doi.org/10.3389/fpsyg.2017.00386

Portnoy, L., & Schrier, K. (2019). Using games to support STEM curiosity, identify, and self-efficacy. *The Journal of Games, Society, and Self, 1*(1), 66–96.

Raabe, I. J., Boda, Z., & Stadtfeld, C. (2019). The social pipeline: How friend influence and peer exposure widen the STEM gender gap. *Sociology of Education, 92*(2), 105–123. https://doi.org/10.1177/0038040718824095

Ramsey, L. R. (2017). Agentic traits are associated with success in science more than communal traits. *Personality and Individual Differences, 106*, 6–9. https://doi.org/10.1016/j.paid.2016.10.017

Ratschinski, G. (2009). *Selbstkonzept und Berufswahl. Eine Überprüfung der Berufswahltheorie von Gottfredson an Sekundarschülern* [Self-concept and career choice. A review of Gottfredson's career choice theory among secondary school students]. Münster: Waxmann.

Reardon, S. F., Kalogrides, D., Fahle, E. M., Podolsky, A., & Zárate, R. C. (2018). The relationship between test item format and gender achievement gaps on math and ELA tests in 4th and 8th grade. *Educational Researcher, 47*(5), 284–294.

Reilly, D. (2012). Gender, culture and sex-typed cognitive abilities. *PLoS ONE, 7*(7), e39904. https://doi.org/10.1371/journal.pone.0039904

Reilly, D., Neumann, D. L., & Andrews, G. (2019). Investigating gender differences in mathematics and science: Results from the 2011 trends in mathematics and science survey. *Research in Science Education, 49*(1), 25–50.

Renfrow, D. G., & Howard, J. A. (2013). Social psychology of gender and race. In J. DeLamater, & A. Ward (Eds.), *Handbook of Social Psychology* (pp. 491–531). Dordrecht: Springer. https://doi.org/10.1007/978-94-007-6772-0_17.

Rice, L., Barth, J. M., Guadagno, R. E., Smith, G. P. A., & McCallum, D. M. (2013). The role of social support in student's perceived abilities and attitudes toward math and science. *Journal of Youth and Adolescence, 42*(7), 1028–1040.

Riegle-Crumb, C., Farkas, G., & Muller, C. (2006). The role of gender and friendship in advanced course taking. *Sociology of Education, 79*, 206–228. https://doi.org/10.1177/003804070607900302

Riegle-Crumb, C., Moore, C., & Ramos-Wada, A. (2011). Who wants to have a career in science or math? Exploring adolescents' future aspirations by gender and race/ethnicity. *Science Education, 95*(3), 458–476. https://doi.org/10.1002/sce.20431

Rittmayer, A.D., & Beier, M.E. (2009). Overview: Self-Efficacy in STEM. *SWE-AWE CASEE Overviews*. Retrieved from http://www.AWEonline.org.

Robnett, R. D., & Leaper, C. (2013). Friendship groups, personal motivation, and gender in relation to high school students' STEM career interest. *Journal of Research on Adolescence, 23*, 652–664. https://doi.org/10.1111/jora.12013

Robnett, R. D., & Thoman, S. E. (2017). STEM success expectancies and achievement among women in STEM majors. *Journal of Applied Developmental Psychology, 52*, 91–100. https://doi.org/10.1016/j.appdev.2017.07.003

Roche, R., & Manzi, J. (2019). Bridging the confidence gap: Raising self-efficacy amongst urban high school girls through STEM education. *American Journal of Biomedical Science and Research, 5*(6), 452–454. https://doi.org/10.34297/AJBSR.2019.05.000964.

Sadker, D., Sadker, M., & Zittleman, K. R. (2009). *Still failing at fairness: How gender bias cheats girls and boys in school and what we can do about it*. Simon & Schuster Inc.

Sadler, P. M., Sonnert, G., Hazari, Z., & Tai, R. (2012). Stability and volatility of STEM career interest in high school: A gender study. *Science Education, 96*, 411–427. https://doi.org/10.1002/sce.21007

Sahranavard, M., Hassan, S., Elias, H., & Abdullah, M. (2012). Student's psychological factors and science performance: Does gender matter for Iranian students. *Life Science Journal, 9*(3), 2069–2075.

Sax, L. J., Kanny, M. A., Riggers-Piehl, T. A., Whang, H., & Paulson, L. N. (2015). "But I'm not good at math": The changing salience of mathematical self-concept in shaping women's and men's STEM aspirations. *Research in Higher Education, 56*, 813–842. https://doi.org/10.1007/s11162-015-9375-x

Schillinger, F. L., Mosbacher, J. A., Brunner, C., Vogel, S. E., & Grabner, R. H. (2021). Revisiting the role of worries in explaining the link between test anxiety and test performance. *Educational Psychology Review, 33*, 1887–1906. https://doi.org/10.1007/s10648-021-09601-0

Schmader, T., Johns, M., & Barquissau, M. (2004). The costs of accepting gender differences: The role of stereotype endorsement in women's experience in the math domain. *Sex Roles, 50*, 835–850. https://doi.org/10.1023/B:SERS.0000029101.74557.a0

Schneider, B., Krajcik, J., Lavonen, J., Salmela-Aro, K., Broda, M., Spicer, J., Bruner, J., Moeller, J., Linnansaari, J., Juuti, K., & Viljaranta, J. (2016). Investigating optimal learning moments in U.S. and Finnish science classes. *Journal of Research in Science Teaching, 53*, 400–421. https://doi.org/10.1002/tea.21306

Sjoberg, S. (1988). Gender and the image of science. *Scandinavian Journal of Education Research, 32*, 49–60.

Smeding, A. (2012). Women in science, technology, engineering, and mathematics (STEM): An investigation of their implicit gender stereotypes and stereotypes' connectedness to math performance. *Sex Roles, 67*(11–12), 617–629. https://doi.org/10.1007/s11199-012-0209-4

Smyth, F. L., & Nosek, B. A. (2015). On the gender-science stereotypes held by scientists: Explicit accord with gender-ratios, implicit accord with scientific identity. *Frontiers in Psychology, 6*, 415. https://doi.org/10.3389/fpsyg.2015.00415

Sobieraj, S., & Krämer, N. C. (2019). The impacts of gender and subject on experience of competence and autonomy in STEM. *Frontiers in Psychology, 10*, 1432. https://doi.org/10.3389/fpsyg.2019.01432

Spencer, S. J., Steele, C. M., & Quinn, D. M. (1999). Stereotype threat and women's math performance. *Journal of Experimental Social Psychology, 35*, 4–28. https://doi.org/10.1006/jesp.1998.1373

Steele, C. M. (1997). A threat in the air: How stereotypes shape intellectual identity and performance. *American Psychologist, 52*, 613–629. https://doi.org/10.1037/0003-066X.52.6.613

Steele, J. (2003). Children's gender stereotypes about math: The role of stereotype stratification. *Journal of Applied Social Psychology, 33*, 2587–2606. https://doi.org/10.1111/j.1559-1816.2003.tb02782.x

Steffens, M. C., & Jelenec, P. (2011). Separating implicit gender stereotypes regarding math and language: Implicit ability stereotypes are self-serving for boys and men, but not for girls and women. *Sex Roles, 64*(5–6), 324–335. https://doi.org/10.1007/s11199-010-9924-x

Steffens, M. C., Jelenec, P., & Noack, P. (2010). On the leaky math pipeline: Comparing implicit math-gender stereotypes and math withdrawal in female and male children and adolescents. *Journal of Educational Psychology, 102*(4), 947–963. https://doi.org/10.1037/a0019920

Steinke, J., & Tavarez, P. M. P. (2018). Cultural representations of gender and STEM: Portrayals of female STEM characters in popular films 2002–2014. *International Journal of Gender, Science and Technology, 9*, 244–277.

Stout, J. G., Dasgupta, N., Hunsinger, M., & McManus, M. A. (2011). STEMing the tide: Using in-group experts to inoculate women's self-concept in science, technology, engineering, and mathematics (STEM). *Journal of Personality and Social Psychology, 100*, 255–270. https://doi.org/10.1037/a0021385

Stout, J. G., Ito, T. A., Finkelstein, N. D., & Pollock, S. J. (2013). How a gender gap in belonging contributes to the gender gap in physics participation. *AIP Conference Proceedings, 1513*(1), 402–405.

Su, R., & Rounds, J. (2015). All STEM fields are not created equal: People and things interests explain gender disparities across STEM fields. *Frontier in Psychology, 6*, 189. https://doi.org/10.3389/fpsyg.2015.00189.

The Lyda Hill Foundation & The Geena Davis Institute On Gender In Media. (2018). *Portray her: Representation of women STEM characters in media.* Geena Davis Institute on Gender in Media, The Lyda Hill Foundation.

Unger, R. K. (1979). Toward a redefinition of sex and gender. *The American Psychologist, 34*, 1085–1094. https://doi.org/10.1037/0003-066X.34.11.1085

Vygotsky, L. S. (1980). *Mind in society: The development of higher psychological processes.* Harvard University Press.
Walton, G. M., Cohen, G. L., Cwir, D., & Spencer, S. J. (2012). Mere belonging: The power of social connections. *Journal of Personality and Social Psychology, 102*, 513–532. https://doi.org/10.1037/a0025731
Wang, M.-T., Eccles, J. S., & Kenny, S. (2013). Not lack of ability but more choice: Individual and gender differences in choice of careers in science, technology, engineering, and mathematics. *Psychological Science, 24*(5), 770–775. https://doi.org/10.1177/0956797612458937
Wang, J., Oliver, J. S., & Staver, J. R. (2008). Self-concept and science achievement: Investigating a reciprocal relation model across the gender classification in a crosscultural context. *Journal of Research in Science Teaching, 45*(6), 711–725. https://doi.org/10.1002/tea.20182
Weinburgh, M. (1995). Gender difference in student attitudes toward science: A meta-analysis of the literature from 1970 to 1991. *Journal of Research in Science Teaching, 32*, 387–398.
Weinreich-Haste, H. (1981). The image of science. In A. Kelly (Ed.), *The missing half: Girls and science education* (pp. 216–229). Manchester University Press.
Woolley, A. W., Chabris, C. F., Pentland, A., Hashmi, N., & Malone, T. W. (2010). Evidence for a collective intelligence factor in the performance of human groups. *Science, 330*, 686–688. https://doi.org/10.1126/science.1193147
Worell, J. (2001). *Encyclopaedia of women and gender: Sex similarities and differences and the impact of society on gender*. Academic Press.
World Economic Forum. (2017). *The Global Gender Gap Report.* WEF. Retrieved November 11, 2018, from www.weforum.org/docs/WEF_GGGR_2017.pdf.
Zeldin, A. L., Britner, S. L., & Pajares, F. (2008). A comparative study of the self-efficacy beliefs of successful men and women in mathematics, science, and technology careers. *Journal of Research in Science Teaching, 45*(9), 1036–1058. https://doi.org/10.1002/tea.20195
Zeldin, A. L., & Pajares, F. (2000). Against the odds: Self-efficacy beliefs of women in mathematical, scientific, and technological careers. *American Educational Research Journal, 37*, 215–246.
Zimmerman, B. J. (2000). Self-efficacy: An essential motive to learn. *Contemporary Educational Psychology, 25*, 82–91.

Dr. Melissa McMinn has worked in in-service and pre-service teacher education for 15 years, and in post-graduate education for eight years. She is currently facilitating a new suite of initial education teacher programs for the Open Polytechnic in New Zealand. She holds a Master's of Education and a Doctorate of Philosophy in Mathematics and Science Education and has achieved the status of Senior Fellow (SFHEA) in recognition of her teaching and learning support in higher education. She is an active researcher and over the past decade has led projects in mathematics and science anxiety in university students and teachers, learning environments research, and co-researched a number of projects including pedagogy in higher education, and children's perceptions of science and technology, among others.

Chapter 2
Epistemological Beliefs About Science and Their Relations to Gender, Attitudes to Science and Science Achievement in UAE Schools

Dean Cairns

Abstract Beliefs about knowledge and knowing are referred to as epistemological beliefs. Scientific epistemological beliefs relate to the nature and acquisition of science knowledge and, are central to promoting deeper, meaningful learning. As gender performance gaps are evident for science achievement in UAE government schools, this study seeks to investigate the relative importance of science epistemological beliefs for 15-year-olds using data drawn from the PISA 2015 cycle. Results of mediation analyses for the whole group model, indicate that overall enjoyment of science and instrumental motivation to learn science are the strongest predictors of scientific literacy and these effects are partially mediated by students' epistemological beliefs about science. Statistically significant differences between the male-only and female-only groups were observed for the hypothesised models employed in the study. For the male-only group, epistemological beliefs about science negated the negative relationship between science achievement and interest in science, and the direct effect of extrinsic motivation on science achievement was stronger than for females. These findings provide evidence relating to the importance of scientific epistemological beliefs for males *and* females and their associated peripheral attitudes and beliefs about learning. This chapter outlines the practical implications for science teaching methodologies in UAE schools.

Keywords Epistemological beliefs · Scientific literacy · Attitudes to science · Gender

Introduction

The importance of epistemological beliefs about science in science education remains a significant area of study (see, Burbules & Linn, 1991; Elder, 2002; Hofer, 2000, 2004a, b, 2008, 2020; Hofer & Pintrich, 1997; Mason et al., 2013; Muis et al., 2015; Sengul et al., 2020; Songer & Linn, 1991; Zhao et al., 2021). Of particular interest are

D. Cairns (✉)
Emirates College for Advanced Education, Abu Dhabi, UAE
e-mail: dcairns@ecae.ac.ae

M. Dickson et al. (eds.), *Gender in STEM Education in the Arab Gulf Countries*,
https://doi.org/10.1007/978-981-19-9135-6_2

the observed relationships between an individual's beliefs about science knowledge (the nature of science knowledge and justifications for its validity) and their beliefs about learning. For example, existing research has established links between the level of sophistication of an individual's epistemological beliefs about science and other learning-related concepts such as an increased motivation to learn (Mason et al., 2013) and improved conceptual understanding (Mason et al., 2008; Songer & Linn, 1991). These relationships are hypothesised to be driven by the theory-like nature of individual epistemological belief systems (Hofer & Pintrich, 1997). The gendered nature of epistemological beliefs about science has not been investigated as thoroughly and, to date, limited evidence exists as to whether gender influences an individual's beliefs about science knowledge (Conley et al., 2004).

This chapter aims to investigate the relative importance of epistemological beliefs about science for males and females in the UAE government school system with a focus on the relationship with science achievement. Epistemological beliefs about science are likely related to student dispositions to learning science (as precursors to developing beliefs) and these constructs will be included in the study, with the intention of providing practical implications to support the development of students' epistemological beliefs about science in UAE classrooms.

Epistemology

The field of epistemology is primarily concerned with answering questions such as "What do we know?" and "How do we know what we know?" The type of knowledge of interest in epistemology is propositional knowledge (knowledge *that* something is true) rather than say, procedural knowledge (knowledge of *how* to do something; Fumerton, 2009; Goldman & McGrath, 2015). As such, epistemologists have traditionally been interested in whether beliefs about the propositions (for example, "I know that the Earth is spherical") are justified or warranted by seeking to discover or invent proper methods of inquiry and investigation. Words such as "justified" and "warranted" are evaluative terms, so advocacy for these methods is a normative activity—in other words, epistemologists are promoting good practice as the only permissible route to justifying knowledge as true. Thus, epistemology can be described as the study of better versus worse ways to think, reason and form opinions such that these views are rational and justified (Goldman & McGrath, 2015). The objects under evaluation are often then, arguments or forms of inference and, being propositions, they can be evaluated using deductive or inductive logic. Inferences of interest are the processes of belief formation (or revision) as a sequence of psychological states (and includes processes such as perception, memory, and problem-solving). Essentially, a strong justification for a proposition means that is it more likely to be true. Thus, interest in these cognitive processes is related to epistemic justification and the evaluation of beliefs—or more accurately evaluation of the belief-forming *processes*. These processes are supported, in part, by tools and instruments (such as the methodologies of empirical science) and thus, epistemology relates also to the evaluation of these instruments.

Beliefs are often conceptualised as binary, one either believes something or not. However, there are usually degrees of conviction or confidence for any given assertion (the probability that something is true). This belief is dependent on the mental state of the knower in relation to the proposition (propositional attitudes). Forms of propositional attitudes include conative or optative attitudes towards a said proposition. Conative attitudes guide the belief in the proposition (either in favour or in opposition) whereas optative attitudes are based purely on an intellectual assessment. Or more simply, the belief (and level of certainty of this belief) are independent of whether one wishes something is true or not. In this sense, to have knowledge you must believe it with a relatively high degree of confidence, and it must be justifiable through accepted (normative) processes (Fumerton, 2009; Goldman & McGrath, 2015). However, it could be argued that an individual cannot *decide* to believe something, but it is possible for said individual to perform actions that increase the likelihood of holding a belief (Fumerton, 2009). It follows then, that if an individual can be guided to perform actions (say, in a science classroom) that improve their chances of believing (true) science knowledge through the application of normative processes for generating science knowledge, then surely teachers play a key role in developing students' personal epistemologies.

Science Epistemology

The requirements for the belief that a claim is very likely to be true, in the field of science, requires a focus on the processes of justification. The traditional view of science education as being a body of facts that must be related to the learners does not typically convey the philosophy of science as a way of knowing. The processes that form this way of knowing must also be learned and understood (Wenning, 2009). Scientific knowledge is gained through three main pathways (to a greater or lesser extent). Firstly, rationalism is the application of logic, based on premises, in determining the truth of a claim. However, scientists seldom rely on this approach as the premises, if not tested, may be false and therefore lead to false logical conclusions (for example, Aristotle claimed that heavy objects fall fasters than lighter objects, based on reason alone). Coherentism is essentially the requirement of consensus through the idea of credible authority, however, this path can lead to inconsistencies, as agreement, even with the acceptance of a credible authority, does not necessarily make a claim true. Particularly in relation to scientific knowledge, this approach could be a barrier to the acceptance of new knowledge that conflicts with what we "know" to be true (e.g. Einstein's theory of relativity). The final pathway, and one most consistently applied in the sciences, is empiricism. In this case, logic is combined with a process of verification (through observation) to generate knowledge. As such, reason is *constrained* by physical evidence. Of course, observations are subject to individual perspectives and biases, but technology and carefully applied scientific processes have allowed for increased precision and reliability in observations (or measurements). Even so, scientific knowledge remains tentative. A scientific fact that we know (and that is accepted as the scientific consensus by scientific institutions) can still be overturned by new empirical evidence. However, although scientific knowledge is not fixed, the vast majority of this knowledge (particularly at the school level) is very unlikely to change. For this reason, scientific knowledge can be described as tentative but also durable (Wenning, 2009).

Epistemological Beliefs

Personal epistemologies (or epistemological beliefs) refer to an individual's beliefs about the nature of knowledge (such as the certainty of knowledge or how it is conceptualised) and the processes by which it is justified or developed. According to Rokeach (1972) beliefs are organised into an architectural system, are describable and measurable, and consequently lead to observable behaviours. The assumption here is that not all beliefs are equal and that some beliefs are more central while others exist in a less-central (peripheral) dimension. Such central "core" beliefs are more resistant to change but, when such a belief does change, the repercussions on the remaining belief system are more significant.

Rokeach (1972) theorised that important, central beliefs tended to be highly connected with other beliefs. This functional connectedness was assumed to rank from highest to lowest for (1) existential beliefs, (2) shared beliefs about existence, (3) derived beliefs and (4) beliefs concerning matters of taste. Existential beliefs relate directly to a person's physical and social world and consequently their identity, therefore they are assumed to have a high number of connections and consequences for other beliefs. These types of beliefs may be shared (or not shared) with others, those that are shared are, again, considered to have a high degree of connectedness. Derived beliefs stem from indirect influence by reference persons or groups rather than from direct encounters with the object of belief. Derived beliefs are assumed to have fewer connections (and consequences) than the beliefs from which they are derived. Lastly, beliefs about matters of taste are largely perceived as arbitrary and have relatively few relations to other beliefs.

This reasoning was used to derive five belief types that ranged along the continuum of core beliefs to beliefs that are more peripheral in nature. The beliefs ranged from Type A beliefs that are fundamental to our psychological existence (and thus functionally related to many other beliefs) through to Type E beliefs about personal tastes (that are generally not related to other beliefs). According to Brownlee et al. (2002), epistemological beliefs can be described as Type C (authority) beliefs. Such beliefs are derived from a more critical view of authorities as both a positive and negative source of evidence for certain beliefs. Type C beliefs are therefore generally resistant to change but they are more changeable than the core, primitive Type A and Type B beliefs. Following this, beliefs about learning are then described as derived

(Type D) beliefs that are underpinned by epistemological beliefs (Type C beliefs; Brownlee et al., 2002).

Although the concept of the centrality of epistemological beliefs is discussed by Hofer and Pintrich (1997), this element is referring to the conceptually "core" elements of what should be considered critical dimensions when theorising about epistemological beliefs. For example, Hofer and Pintrich (1997) analysed six theories of epistemological beliefs and thinking, and categorised components of each of the six models as either core dimensions of epistemological theories or peripheral beliefs about learning, instruction, and intelligence. The peripheral natures of these dimensions were related to their lack of *direct* connectedness with the concept of epistemology. In other words, to maintain conceptual clarity, measures of epistemological beliefs were limited to beliefs about the nature of knowledge and the processes of knowing (Hofer & Pintrich, 1997). Individuals' epistemological beliefs are described as theory-like (consisting of the elements of coherence, ontological distinctions and a limited causal-explanatory framework) and, as such, may well function to influence individual thinking in other processes (such as learning). However, the researchers also point out epistemological thinking develops quite late in a child's development and younger children are more likely to form beliefs relating to their experiences of teaching and learning *prior* to developing beliefs about knowledge itself. These peripheral beliefs about learning may be developmental precursors to their core ideas about epistemology.

Not surprisingly, epistemological beliefs remain a prominent area of study in educational research (see Mason et al., 2013; Muis et al., 2015). The importance of epistemological beliefs, particularly in terms of cognitively oriented research, stems from the relationship between how an individual interprets and values new knowledge and the process of learning, in fields such as science education (Burbules & Linn, 1991; Songer & Linn, 1991). This association between epistemological beliefs and learning behaviours has been reported in terms of learning strategies (Schommer & Walker, 1995), achievement (Trautwein & Lüdtke, 2007), conceptual understanding (Songer & Linn, 1991) and motivation (Mason et al., 2013).

Epistemological Beliefs About Science

The current understanding of science epistemology encompasses an epistemological beliefs construct that consists of two central domains: the *nature of knowledge* (what is knowledge?), and the *nature or process of knowing* (how do we know?). The two areas of epistemological beliefs within the scientific knowledge domain are theorised to consist of four dimensions (Hofer & Pintrich, 1997; Hofer, 2000, 2004a). Two of these dimensions are associated with the nature of knowledge and they consist of concepts relating to the *simplicity* of science knowledge (ranging from beliefs about scientific knowledge as isolated facts to science knowledge consisting of interrelated, multifaceted information), and the *certainty* of scientific knowledge (which concerns beliefs about whether knowledge is fixed and unchanging or if there is a degree of evidence-informed progression). The remaining two dimensions are linked to the nature of the processes of knowing. The dimensions are the *source* of scientific knowledge (whether one sees themselves as a locus for knowledge generation or if knowledge is solely obtainable from external sources of authority), and *justification* of scientific knowledge (beliefs relating to whether scientific knowledge can be justified as true based on what an individual feels is correct, through to rigorous, well-established, inquiry processes being used to test the validity of knowledge claims).

The importance of sophisticated epistemological beliefs for science learning has been explored in various contexts. For example, Mason et al. (2008) demonstrated a link between beliefs relating to the continuously evolving nature of scientific knowledge and improved conceptual knowledge development. A study relating to evolution as a theory reported that epistemological beliefs (in this case that scientific knowledge is constantly changing based on evidence) were related to the acceptance of the theory of human evolution (Sinatra et al., 2003). This effect was only observable for potentially controversial theories (for example there was no such relationship found when studying the theory of animal evolution). Another study, involving 212 grade 9–12 students in Georgia by Qian and Alvermann (1995), revealed that the naïve belief that scientific knowledge is simple and certain was negatively correlated to conceptual change learning. Similar findings were reported for domain-specific academic performance for the dimension of certainty and simplicity for first-year college students (Hofer, 2000). As such, there is good evidence in the existing body of research that advanced epistemological beliefs about science are associated with improved academic achievement.

Gender and Epistemological Beliefs About Science

Studies of epistemological beliefs about science have not uncovered any significant (or convincing) gender effects. For example, Conley et al. (2004) tested for the moderating effects of gender and ethnicity of 5th-grade students and found no significant main effects or interactions relating to gender and any of the belief measures used. An earlier study by Schommer (1993) showed that, for high school students, females tended to have more sophisticated epistemological beliefs than males, for two of the dimensions measured (quick learning and fixed ability). However, these dimensions are more closely related to an individual's beliefs relating to learning, instruction and intelligence than to theories of epistemology. In other words, although gender differences are observed, the constructs under study do not align with the central concepts of the nature of knowledge or the nature of knowing that encapsulate epistemological beliefs (Hofer & Pintrich, 1997). Another study conducted in elementary schools in Turkey measured scientific epistemological beliefs using an instrument consisting of two dimensions that were described as "tentative" and "fixed", with 8 items per dimension (Ozkal et al., 2010). The description of these domains indicated that this instrument focused only on one aspect of the nature of knowledge (the certainty of knowledge) and did not appear to capture beliefs relating to the interrelated nature of knowledge, or beliefs relating to the processes by which one comes to know scientific knowledge, such as the source of the knowledge or how knowledge claims are evaluated (justification). Again, although gender differences are observed the results are inconclusive as the study did not appear to measure both well-established components of the epistemological beliefs construct.

Peripheral Beliefs About Learning Science

In terms of the belief categorisation process described by Rokeach (1972), beliefs about learning that are possibly derived from more central beliefs about knowledge (epistemological beliefs) can be located as peripheral beliefs (Brownlee et al., 2002). Or, at least, these beliefs are less central to an individual's identity and therefore less connected with their overarching belief system than epistemological beliefs. However, Hofer and Pintrich (1997) suggested that epistemological beliefs are possibly formed through earlier teaching and learning experiences (such as enjoyable science lessons) during childhood. As such, although beliefs about learning may be derived from epistemological beliefs, an individual's attitudes and beliefs about learning (generated through direct experience of learning experiences as a young child) may pre-empt the formation of epistemological beliefs. This concept seems particularly salient to the present study relating to the attitudes and beliefs of adolescents—young adults who may have developed their core personal epistemologies from teaching and learning experiences throughout their school life.

Such beliefs about learning include dispositions to learning science such as interest in science, enjoyment of learning science, and science self-efficacy. These dispositions could influence epistemological beliefs and, in turn, science achievement (as will be investigated in this study) or they could directly influence student learning. Certainly, there are a number of studies that associate students' motivations to learn science and their observed learning behaviours (see Bryan et al., 2011; Singh et al., 2002). According to Bryan et al. (2011) motivation can be described as consisting of intrinsic motivation, self-efficacy and self-determination as described in social cognitive theory. Motivation directs and maintains positive goal-oriented learning behaviours such as engagement and participation in lessons and other learning-related activities (Schunk et al., 2014).

Intrinsic motivation can be described as learning behaviours that lead to internal rewards, such as pleasure or satisfaction, which originate from the innate psychological requirements of competence and self-determinism (Vallerand et al., 1992). Gottfried (1985) reported that academic intrinsic motivation involving junior high school students was, to some extent, domain-specific (including science subjects) and positively and significantly correlated with attainment. In a synthesis, employing the Self-Determination Theory perspective, of four studies (a meta-analysis and three controlled, longitudinal studies involving high school and college students in Sweden and Canada), Taylor et al. (2014) demonstrated that intrinsic motivation consistently predicted science attainment. They went on to claim that intrinsic motivation is the most important "motivational ingredient" for improved student outcomes. In a study by Nugent et al. (2015) investigating STEM learning and career orientation, social cognitive career theory was used to develop a structural equation model. They found that an inherent interest in science (a trait that is influenced by educators, friends, and family) positively affected science performance via self-efficacy.

Self-efficacy is described as the perceived effectiveness of one's actions in dealing with various situations (Bandura, 1982) and science self-efficacy has been found to be a strong predictor of science achievement (Britner & Pajares, 2006). For example, science self-efficacy has been positively correlated with GPA scores for 281 grade 7 students in school science classes (Pajares et al., 2000) and there is also evidence for a mediation effect of self-efficacy on other beneficial learning skills or beliefs by influencing behaviours such as effort and persistence, irrespective of cognitive ability (Bouffard-Bouchard et al., 1991).

Extrinsic motivation is related to learning behaviours that are carried out in order to achieve future goals (namely, how useful the knowledge or qualification may be in the future) and is not related to interest in the activities themselves (Vallerand et al., 1992). Extrinsic motivation, from a teaching and learning perspective, has been described as contrasting with intrinsic motivation (Harter, 1981) and even undermining intrinsic interest (Deci, 1971). Harter (1981) also concedes that there could be scenarios where extrinsic and intrinsic motivation could cooperate and lead to improved learning behaviours. For example, Deci (1971) argues that extrinsic motivation is affected by the student's environment and is therefore adaptable (and not trait-like) leading to possible situations where an intrinsically motivated learner is further motivated by external rewards in the classroom. For example, a learner may engage in a learning activity because it interests them *and* because it gains approval from their teacher, or a learner may be an independent problem solver that requires teacher guidance later in the learning process (Lepper et al., 2005). Although in terms of college-level students, increased extrinsic motivation can have a positive impact on attainment by achieving learning goals (Barron & Harackiewicz, 2001), negative learning behaviours, especially in school-level learners have also been observed (Ames & Archer, 1988; Dweck, 2013; Meece et al., 1988). An extrinsic motivation focus (in these cases a focus on performance goals, pleasing the teacher, and gaining social recognition) was reported to lead to learning behaviours such as the avoidance of challenge (Dweck, 2013), associating failure with a lack of innate ability (Ames & Archer, 1988), and lower levels of cognitive engagement in science classes (Meece et al., 1988).

Present Study

The present study seeks to determine the relationships between epistemological beliefs about science and 15-year-old male and female students' attitudes to learning science as possible precursors to these beliefs. It is hypothesised that their attitudes and beliefs about learning science, developed during their school science experiences, influence their epistemological beliefs about science and, in turn, their science achievement. Additionally, we suggest that the hypothesised model (see Fig. 2.1) should fit the data for both female and male students as no convincing gender differences in overall personal epistemologies have been reported in the literature. This study then investigates the following two research questions:

1. Are the relationships between attitudes and beliefs about learning science and science achievement mediated by epistemological beliefs about science?
2. Do the relationships between attitudes and beliefs about science learning, epistemological beliefs about science, and science achievement differ for males and females?

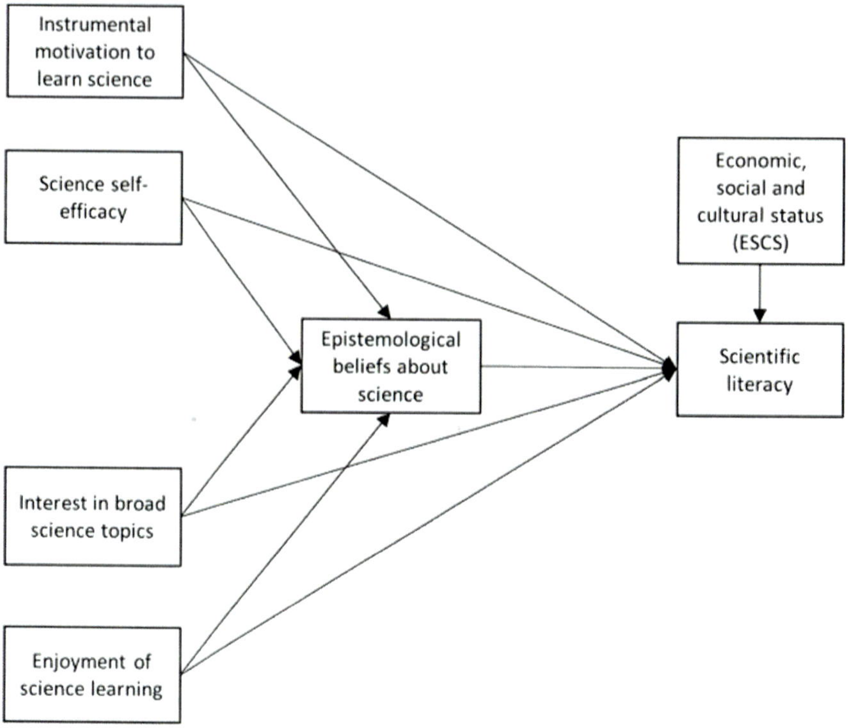

Fig. 2.1 Hypothesised model

Methodology

Data

The goal of the Programme for International Student Assessment (PISA), an initiative of the Organisation for Economic Cooperation and Development (OECD) in Paris, is to provide reliable information regarding educational achievement across countries (Thomson & De Bortoli, 2008). The participants in this study are 15-year-old students that took part in the 2015 PISA cycle attending government-funded schools in the UAE. The sample data was drawn from the published data for the UAE (https://www.oecd.org/pisa/data/2015database/) and filtered using the school ownership variable such that only government-owned schools were included in the analysis. The resulting dataset consisted of 5158 students from 470 schools across all 7 Emirates. The gender breakdown of this sample included 2440 males (47% of the sample) and 2718 females (53% of the sample).

Measures

Epistemological Beliefs About Science

Self-reported epistemological beliefs about science were determined using the epistemological belief index in the PISA 2015 student questionnaire (OECD, 2016, p. 241) where students were asked to respond to the following statements: "A good way to know if something is true is to do an experiment" (*nature or process of knowing*); "Ideas in science sometimes change" (*nature of knowledge*); "Good answers are based on evidence from many different experiments" (*nature or process of knowing*); "It is good to try experiments more than once to make sure of your findings" (*nature or process of knowing*); "Sometimes scientists change their minds about what is true in science" (*nature of knowledge*); "The ideas in science books sometimes change" (*nature of knowledge*). The reliability of the scale in the UAE sample (all schools) was shown to be good (Cronbach's alpha = 0.874).

Scientific Literacy

Science achievement is referred to as scientific literacy in this study. This is the term used by the OECD in the PISA 2015 cycle to refer to the assessment of knowledge about science and technology, and students' understanding of investigative processes used during scientific inquiry (OECD, 2017). One of the 3 areas of competency (that is particularly relevant to this study) within the scientific literacy construct is epistemic knowledge. This is measured by items that question students' knowledge of science-specific constructs (e.g. theories, models, laws, observations, and hypotheses), and the roles of these constructs in justifying scientific knowledge (e.g. how claims about scientific knowledge are supported by data and reasoning, the importance of measurement error, and the purpose of the peer review process). Further details relating to the design, implementation and standardisation processes employed to measure scientific literacy can be found in the published Assessment Framework (OECD, 2017).

Enjoyment of Science

In terms of the PISA scales employed in this study, intrinsic motivation is measured using the "Enjoyment of science" scale where students were asked for their level of agreement (strongly disagree to strongly agree) with statements relating to their enjoyment of doing and learning science (OECD, 2016). Items included in the scale were "I generally have fun when I am learning science topics", "I like reading about science", "I am happy working on science topics", "I enjoy acquiring new knowledge in science", and "I am interested in learning about science". The reported item parameters for this scale derived from the item response theory analysis (employing the generalised partial credit model) showed that the items were of a similar level of discrimination (alpha ranged from 0.83 to 1.15) meaning that each item contributed approximately the same level of information to the model and the resulting weighted likelihood estimates used to indicate the scale score for each student's level of enjoyment of science. The reliability of the scale in the UAE sample was reported as very good (Cronbach's alpha = 0.929).

Science Self-efficacy

Self-efficacy is measured in PISA 2015 using student responses to items that question their perceived ability to understand and analyse real-world situations using their current level of science knowledge (OECD, 2016). Respondents were asked how easy it would be for them to perform science-related tasks by responding on a four-point scale consisting of the categories "I could do this easily", "I could do this with a bit of effort", "I would struggle to do this on my own" and "I couldn't do this". A sample of tasks included in the scale was to "Recognise the science question that underlies a newspaper report on a health issue", "Describe the role of antibiotics in the treatment of disease" and "Identify the science question associated with the disposal of garbage". Again, all items provided a similar level of discrimination and contributed equally to the scale score. Furthermore, the reliability of the scale in the UAE sample was reported as good (Cronbach's alpha = 0.886).

Instrumental Motivation to Learn Science

In PISA 2015, extrinsic motivation to learn science was determined using the instrumental motivation scale where students responded to questions relating to their future career or study plans (OECD, 2016). Again, the students were asked to provide their level of agreement with a range of statements on a four-point scale. The statements in the scale were "Making an effort in science subject(s) is worth it because this will help me in the work I want to do later on", "What I learn in my science subject(s) is important for me because I need this for what I want to do later on", "Studying science subject(s) is worthwhile for me because what I learn will improve my career prospects" and "Many things I learn in science subject(s) will help me to get a job". Again, this scale was shown to be reliable in the UAE sample (Cronbach's alpha = 0.899).

Control Variables

To allow for factors that are likely to affect the other variables in the study (that are not related to their school science experience) students' economic, and cultural status (ESCS) was included in the model as a control variable. No income-related data is collected as part of the PISA data collection process, so students' ESCS is derived from questions relating to the education level and occupation of the respondent's parents and information regarding the availability of various possessions in their home (e.g. availability of books, a location to study, internet access, and the like). The reliability of this composite scale for the UAE sample was very low (Cronbach's alpha = 0.36) and this is due to the low reliability of the underlying home possessions scale. This is likely caused by the comparative wealth of the UAE compared to the average wealth of the participating countries used in the scaling model for the home possessions construct (Avvisati, 2020).

Analytic Approach

Research questions 1 and 2 of this study are addressed through the use of path analysis that is used, in this case, to describe a form of structural equation modelling that does not include a measurement model and, as a result, uses only single indicators in the structural model (Loehlin & Beaujean, 2017). The indicators used are the existing scales as described in the sections above.

The hypothesised model tested in this study is shown in Fig. 2.1.

To address Research Question 1 the hypothesised model (Fig. 2.1) was generated using the data for the whole sample (males and females) and the mediation analysis was conducted using the TYPE = complex command (to account for the sampling methods used during data collection) and the *model indirect* function in Mplus (Muthén & Muthén, 1998–2017) using bootstrapped bias-corrected confidence intervals to account for the lack of normality in mediation effects (Preacher & Hayes, 2008). The goodness-of-fit indices used to test the suitability of the model to describe the relationships between the variables were the Standardised Root Mean Square Residual (SRMR), the Root Mean Square Error Of Approximation (RMSEA), the comparative fit index (CFI), and the Tucker–Lewis index (TFI).

To address Research Question 2, the hypothesised model was tested as a multi-group analysis where group 1 was males and group 2 was females. Again, the previously mentioned goodness-of-fit indices were used to test the model fit when treating males and females as two distinct groups. Further testing of model fit was required to identify differences between males and females. This was tested using the chi-squared difference test whereby the male and female models were compared by allowing the regression coefficients to vary between groups (male and female model regression coefficients were estimated independently) and comparing the model fit with the same models where regression coefficients were constrained across groups (male and female model regression coefficients were estimated as a single value). This more constrained model (M_0) is thus nested within the less constrained, baseline model (M_1). However, the Satorra–Bentler scaled (mean-adjusted) chi-square value (helpfully reported by Mplus when using the MLR estimator) is not directly useable in chi-squared difference testing because this difference is not distributed as chi-square (Bryant & Satorra, 2012). As a result, it is important to divide the reported maximum likelihood chi-square value by the Satorra–Bentler scaled chi-square value to calculate the model's scaling correction factor prior to carrying out the difference test (see https://www.statmodel.com/chidiff.shtml). As with standard methods for chi-squared difference testing, it is now possible to subtract the derived chi-squared value for M_1 from the value for M_0 and test the difference in values (taking into account the difference in degrees of freedom) to determine if the constraints added to M_0 significantly worsen model fit. If, for example, the difference is insignificant then the null hypothesis (that there is no difference between the fit of the models) would be accepted. In such a case it would be possible to conclude that there is not a significant difference between males and females in terms of the relationships depicted in the hypothesised model.

Results

Unweighted descriptive statistics and correlations for the variables included in this model are included in Tables 2.1 and 2.2, respectively. The inter-variable correlations (Table 2.2) show that there are sufficient zero-order relationships between the constructs to justify inclusion in the model.

The results of the path analysis carried out according to the hypothetical model (Fig. 2.1) using the data for males and females as a single group are shown in Fig. 2.2.

Table 2.1 Unweighted descriptive statistics

Variables	N	Mean	SD
Scientific literacy	4360	402.87	82.41
Epistemological beliefs[a]	4325	−0.10	1.04
Enjoyment of science[a]	4360	0.38	1.09
Interest in broad science topics[a]	4360	0.14	1.01
Instrumental motivation to learn science[a]	4360	0.64	0.84
Science self-efficacy[a]	4360	0.34	1.30
ESCS[a]	4360	0.33	0.81
Gender	5158	0.54	–

Note ESCS = economic, social and cultural status
[a]Reported as a weighted likelihood estimate standardised to a mean of 0 and a SD of 1 for OECD countries

Table 2.2 Unweighted Pearson correlation matrix for continuous variables

Variables	a	b	c	d	e	f	g
a. Scientific literacy	–						
b. Epistemological beliefs	0.37	–					
c. Enjoyment of science	0.24	0.39	–				
d. Interest in broad science topics	0.05	0.25	0.44	–			
e. Instrumental motivation	0.26	0.27	0.37	0.20	–		
f. Science self-efficacy	0.14	0.13	0.21	0.14	0.32	–	
g. ESCS	0.12	0.04	0.01	0.05	0.08	0.13	–

Note ESCS = economic, social and cultural status

The goodness-of-fit measures provided in the Mplus output indicate that the hypothesised model imposed on the data fits the data well. Due to the use of the *robust* maximum likelihood estimator, which is required when running complex samples (Muthén & Muthén, 1998–2017), the Chi-Square Test of Model Fit is not reported as the chi-square-difference values are not distributed as chi-square (Byrne, 1996). However, Mplus applies a scaling correction factor (Satorra–Bentler mean-adjusted chi-square) such that this value can be used in the calculation of the more suitable goodness-of-fit indices (Muthén & Muthén, 1998–2017). For example, the chi-squared derived *incremental* fit indices in the output showed good model fit (CFI = 1.000, TLI = 0.996) as both were higher than the recommended cut-off value of 0.95 (Hu & Bentler, 1999). Also reported are the *absolute* indices of fit (or, perhaps more accurately, *misfit* indices; Browne et al., 2002) that do not rely on a comparison with a reference (or baseline) model. These include the Standardised Root Mean Square Residual (SRMR) and the Root Mean Square Error of Approximation (RMSEA). The RMSEA, a measure of fit that is sensitive to the complexity of the model, is reported as 0.009, 90% CI [0.000, 0.042] which again indicates that,

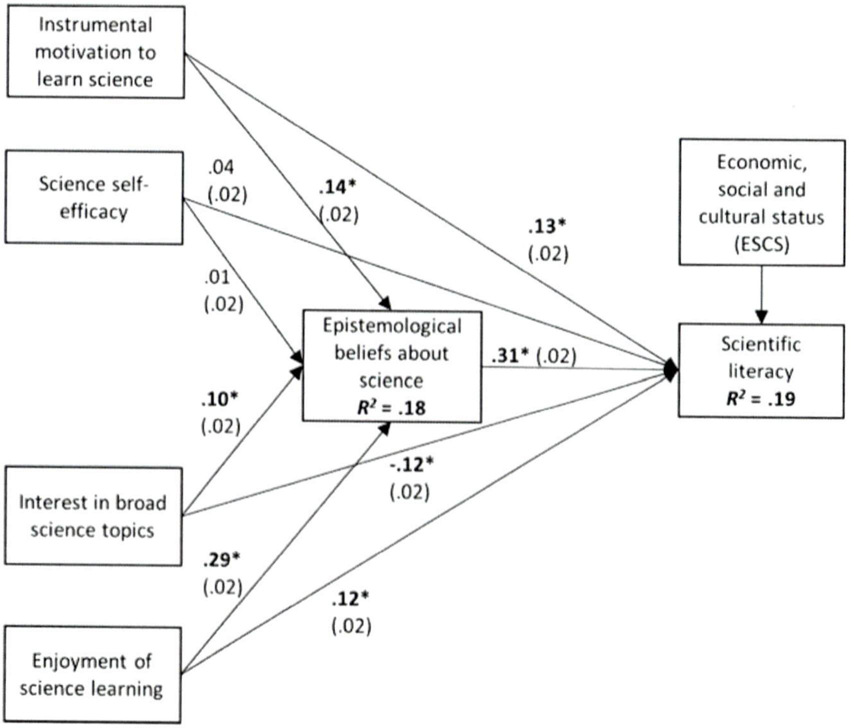

Fig. 2.2 Path analysis results for the whole UAE government school sample. *Note* Standardised regression coefficients reported are for each path. $^*p < 0.01$

even at the upper bounds of the 90% confidence interval, this is a well-fitted model as it is below the suggested cut-off value of 0.06 (Hu & Bentler, 1999). Lastly, the SRMR represents the standardised, average residual value resulting from comparing the variance–covariance matrix of hypothesised model with the matrix for the data, such that the closer the match (i.e. the better the model fits the data) the smaller the value reported. The SRMR for the hypothesised model is 0.006 and, again, indicated a good model fit (Hu & Bentler, 1999).

The results of the path analysis carried out according to the hypothetical model (Fig. 2.1) using the multi-group analysis whereby the sample data was grouped by females and males are shown in Figs. 2.3 and 2.4, respectively.

The goodness-of-fit indices for this multi-group analysis showed that when treating the data of each group as distinct, the hypothesised model (Fig. 2.1) fit the data slightly better, by most measures, than the whole group analysis: CFI = 1.000, TLI = 0.999, RMSEA = 0.005, 90% CI [0.000, 0.043], SRMR = 0.007. Also, the chi-square difference test for the constrained model (regression coefficients held constant between the male and female groups) showed that the model fit between the groups was significantly different ($\Delta MLM_{\chi^2_{[10]}} = 30.080$, $p < 0.001$)

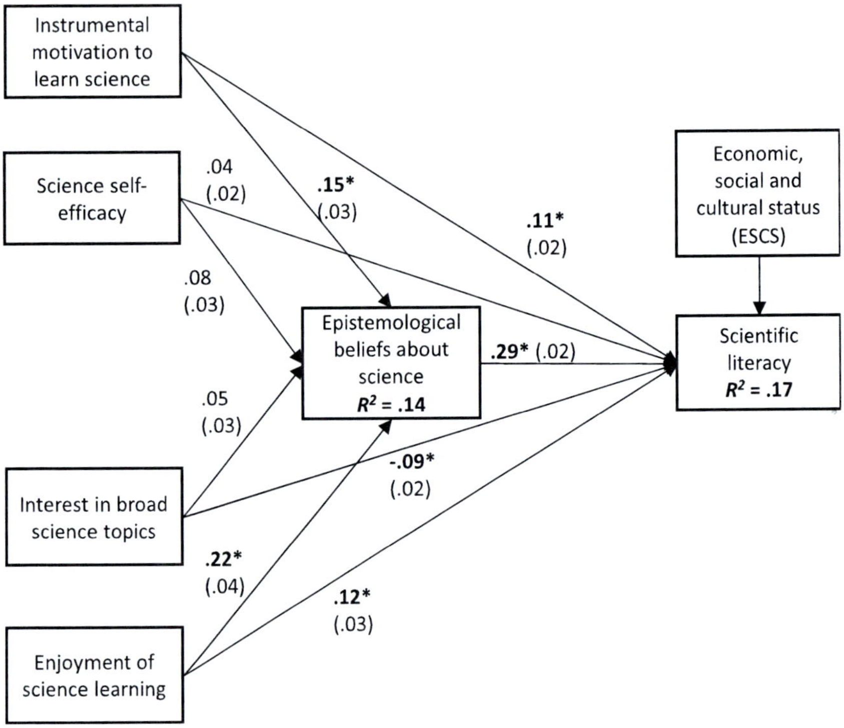

Fig. 2.3 Path analysis results for female students only in UAE government school sample. *Note* Standardised regression coefficients reported for each path. *$p < 0.01$

indicating that the differences observed between the models (in terms of the size and significance of the estimated parameters) were also of statistical significance.

The direct, indirect, and total effects of attitudes to science learning, for each model, are reported in Tables 2.3, 2.4, and 2.5. To improve comparability between groups (males and females) *unstandardised* regression coefficients were reported as recommended (Hayes, 2017). This is because different samples (in this case the male- and female-only samples) have varying standard deviation values (the values used in the standardisation process) but the metric of measurement for the scales and the test score remain the same between the samples. Furthermore, retaining the metric of the test scores enables a more meaningful discussion relating to these variables in terms of absolute test performance and thus the practical significance of the findings.

In the case of the whole group model (Table 2.3) the largest *total* effects were related to the enjoyment of science (estimate = 15.50, 99% CI[12.18, 18.81]) and instrumental motivation to learn science (estimate = 17.12, 99% CI[12.34, 21.90]). Each of these effects was equivalent to approximately 6 months of additional schooling (Jerrim & Shure, 2016). Significant *indirect* effects between students' dispositions to science and science literacy through the mediator of students' epistemological beliefs about science, were observed for the enjoyment of science (estimate

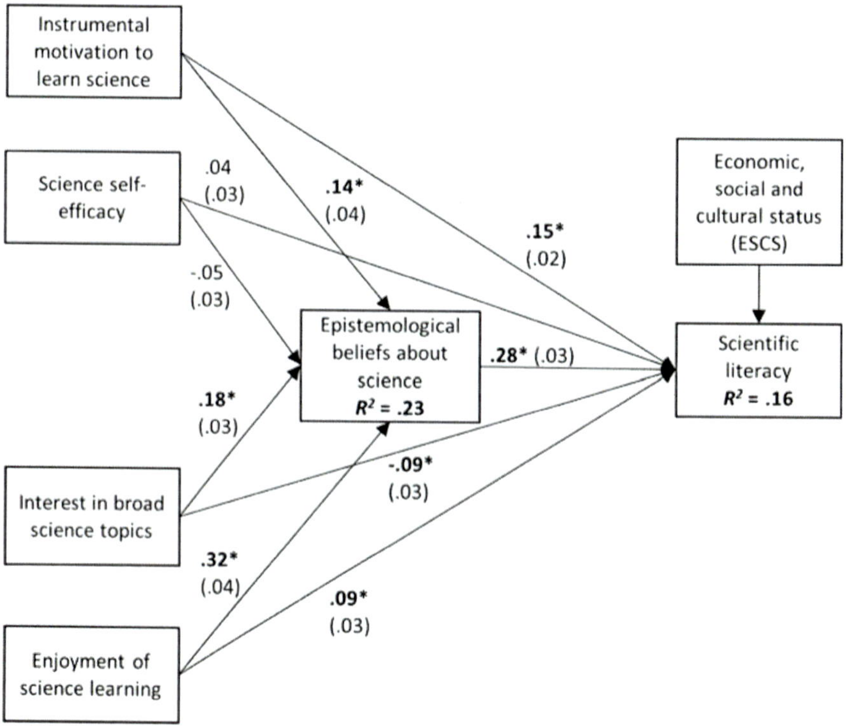

Fig. 2.4 Path analysis results for male students only in the UAE government school sample. *Note* Standardised regression coefficients reported for each path. $*p < 0.01$

= 6.75, 99% CI[4.61, 8.90]), interest in broad science topics (estimate = 2.50, 99% CI[1.08, 3.92]), and instrumental motivation to learn science (estimate = 4.22, 99% CI[2.17, 6.28]).

Table 2.4 shows the mediation analysis results for the female group model. Again, as with the whole group model, the largest *total* effects were concerning the enjoyment of science (estimate = 13.61, 99% CI[9.00, 18.23]) and instrumental motivation to learn science (estimate = 14.09, 99% CI[8.35, 19.83]). In this case, significant *indirect* effects were observed for the enjoyment of science (estimate = 4.66, 99% CI[2.45, 6.87]) and instrumental motivation to learn science (estimate = 4.03, 99% CI[1.50, 6.57]) only.

Table 2.5 represents the results obtained for the male group model. As with the whole group and female-only group models, the largest *total* effects were again related to the enjoyment of science (estimate = 13.55, 99% CI[8.90, 18.19]) and instrumental motivation to learn science (estimate = 18.84, 99% CI[11.21, 26.47]). Significant *indirect* effects were again observed for the enjoyment of science (estimate = 6.62, 99% CI[3.41, 9.82]), interest in broad science topics (estimate = 3.90, 99% CI[1.73, 6.07]), and instrumental motivation to learn science (estimate = 3.83, 99% CI[1.14, 6.52]). In both groups, no significant mediation (or direct) effects were

Table 2.3 Indirect effects of science attitudes to learning on science literacy through students' epistemological beliefs for the whole group using unstandardised regression coefficients

Distal variable	Mediator	Product of coefficients		99% Bootstrap bias-corrected CI	
		Estimate	*SE*	Lower limit	Upper limit
Whole group model					
Enjoyment of science	Epistemological beliefs about science	6.75*	0.83	4.61	8.90
Direct effect		8.74*	1.37	5.20	12.28
Total effect		15.50*	1.29	12.18	18.81
Interest in broad science topics	Epistemological beliefs about science	2.50*	0.55	1.08	3.92
Direct effect		−9.33*	1.57	−13.36	−5.30
Total effect		−6.83*	1.56	−10.84	−2.82
Instrumental motivation	Epistemological beliefs about science	4.22*	0.80	2.17	6.28
Direct effect		12.90*	1.84	8.17	17.63
Total effect		17.12*	1.86	12.34	21.90
Science self-efficacy	Epistemological beliefs about science	0.22	0.43	−0.88	1.31
Direct effect		2.19	1.10	−0.64	5.01
Total effect		2.40	1.19	−0.66	5.46

Note $^*p < 0.01$

detected for science self-efficacy and its relationships with epistemological beliefs about science and science literacy.

Discussion

To address Research Question 1, in the following discussion the findings relating to the full sample (males and females) and the relationships identified between attitudes and beliefs about learning and scientific literacy scores, and the mediating effects of epistemological beliefs about science, will be examined.

The strongest indirect relationship was between the enjoyment of science and scientific literacy, mediated by epistemological beliefs about science. If students report that they generally have fun when learning science, they tend to perform better in the scientific literacy assessment. Although there is a clear direct association between students enjoying science and their performance, the total relationship (an increase of 15.50 points in their scientific literacy score for a 1-point increase in the enjoyment of science) is explained, in part, by improved understanding of how scientific knowledge is generated (epistemological beliefs about science). In other

Table 2.4 Indirect effects of science attitudes to learning on science literacy through students' epistemological beliefs for the female group using unstandardised regression coefficients

Distal variable	Mediator	Product of coefficients		99% Bootstrap bias-corrected CI	
		Estimate	*SE*	Lower limit	Upper limit
Female group model					
Enjoyment of science	Epistemological beliefs about science	4.66*	0.86	2.45	6.87
Direct effect		8.96*	1.88	4.11	13.81
Total effect		13.61*	1.79	9.00	18.23
Interest in broad science topics	Epistemological beliefs about science	1.30	0.65	−0.37	2.96
Direct effect		−7.22*	1.95	−12.25	−2.19
Total effect		−5.92*	1.95	−10.95	−0.90
Instrumental motivation	Epistemological beliefs about science	4.03*	0.98	1.50	6.57
Direct effect		10.05*	2.13	4.57	15.54
Total effect		14.09*	2.23	8.35	19.83
Science self-efficacy	Epistemological beliefs about science	1.40	0.47	−0.20	2.60
Direct effect		1.17	1.50	−2.69	5.04
Total effect		2.57	1.43	−1.12	6.26

Note $*p < 0.01$

words, students who enjoy science tend to also develop more sophisticated understandings of the nature of scientific knowledge and scientific knowledge-building processes (such as methods for the empirical testing of hypotheses) and this appears to support understanding and improve overall scientific literacy. Strong relationships between enjoyment of science and science achievement are well documented (see, Areepattamannil et al., 2011; Cairns & Areepattamannil, 2017) and these findings align with previous research using data from other countries and cross-national samples.

Instrumental motivation to learn science is also positively associated with scientific literacy. That is to say, the more students believe that science will be useful to them in their future studies and careers (as per Expectancy-Value theory; Wigfield & Eccles, 2000) the higher their scores in scientific literacy. Again, this effect is partially mediated by epistemological beliefs about science and some of this positive relationship can thus be described as being transmitted through their improved understanding of the nature and justification of scientific knowledge. This highly positive relationship was largely consistent across countries and economies that participated in PISA 2015 (OECD, 2016) and aligns, overall with the concept of Expectancy-Value Theory, whereby motivation to undertake certain actions (e.g. engaging with science learning) is determined by the perceived likelihood that said action will produce the

Table 2.5 Indirect effects of science attitudes to learning on science literacy through students' epistemological beliefs for the male group

Distal variable	Mediator	Product of coefficients		99% Bootstrap bias-corrected CI	
		Estimate	*SE*	Lower limit	Upper limit
Male group model					
Enjoyment of science	Epistemological beliefs about science	6.62*	1.24	3.41	9.82
Direct effect		6.93*	2.00	1.78	12.08
Total effect		13.55*	1.80	8.90	18.19
Interest in broad science topics	Epistemological beliefs about science	3.90*	0.84	1.73	6.07
Direct effect		−6.57*	2.30	−12.49	−0.65
Total effect		−2.67	2.16	−8.23	2.89
Instrumental motivation	Epistemological beliefs about science	3.83*	1.05	1.14	6.52
Direct effect		15.01*	3.03	7.20	22.82
Total effect		18.84*	2.96	11.21	26.47
Science self-efficacy	Epistemological beliefs about science	−0.84	0.52	−2.19	0.51
Direct effect		2.29	1.70	−2.10	6.67
Total effect		1.45	1.96	−3.59	6.49

Note *$p < 0.01$

desired instrumental outcome (e.g. working in a science-related field), and the overall desire to experience this outcome (Wigfield & Eccles, 2000).

A different relationship was identified for the intrinsic measure of interest (in broad science topics). This is a domain-specific measure and requires the students to express an interest in *something* related to science such as an activity (for instance carrying out experiments), an object (e.g. a microscope), or a field such as forensic science (Krapp & Prenzel, 2011). Interestingly, the overall relationship between interest in broad science topics and scientific literacy is small and negative. The indirect effect, mediated by epistemological beliefs is positive and, to some extent, reduced the negative effect of the interest variable. This suggests that students being interested in science topics without a clear understanding of the nature of scientific knowledge or the methods by which scientific knowledge is justified (empiricism through systematic scientific inquiry) will perform poorly in an assessment where this depth of knowledge is required. In other words, superficial interest in "science" with little understanding of the underlying processes is related to lower levels of scientific literacy.

The final dispositional variable investigated in this model was science self-efficacy. Self-efficacy refers to an individual's beliefs relating to their abilities to perform activities competently and these judgements regarding personal capacity (whether accurate or otherwise) are reflected in the actions of the individual. These could result in avoidance or engagement with said activities (Bandura, 1982). Previous studies have shown that science self-efficacy is related to student science performance for high school students (Burns et al., 2021; OECD, 2016). However, in this study, when accounting for other constructs in the model, science self-efficacy was not related to either scientific literacy or epistemological beliefs about science for students in UAE government schools. This is in direct contrast to prior research relating to the importance of self-efficacy for science learning (Britner & Pajares, 2006; Pajares et al., 2000). This finding is also contrasted with the OECD's analysis of the data from the PISA 2015 assessment, across all participating countries, where a one-point increase on the science self-efficacy scale was associated with a 17-point increase in their science test score (OECD, 2016). However, a closer examination of the OECD analysis on a country-by-country basis, revealed that this relationship was not consistent across all levels of scientific literacy. Students studying in countries with lower overall scientific literacy scores tended to over-estimate their competency regarding the science topics on the scale resulting in a smaller or even negative relationship between self-reported science self-efficacy and test performance (OECD, 2016). Bandura (1997) reported the four main sources of self-efficacy beliefs as *mastery experience* (students' experiences of success in performing a task), *vicarious experiences* (comparison of a student's ability with other student performances), *verbal and social persuasions* (self-efficacy developed through feedback regarding performance from influential others), and *physiological and affective states* (for example, stress and fatigue). More recently, these relationships to self-efficacy were shown to be influenced by students' implicit theories of learning. For instance, a student who believes that their ability is fixed and receives negative feedback (through, say, task performance and teacher feedback) will likely exhibit low self-efficacy (Chen & Usher, 2013). Perhaps, this relationship indicates that students in UAE schools were experiencing a degree of spurious mastery experiences, some erroneous verbal and social feedback, and messaging in their learning environment relating to the fixed nature of their ability in science. These combined phenomena could explain the observed absence of a significant relationship between self-reported science self-efficacy and scientific literacy score in this sample.

Research Question 2 required the results of the male-only and female-only models to be analyzed and compared. Overall, the models for the male group and the female group exhibited a number of similarities in terms of the relationships between the variables in the models. For example, the relationship between science self-efficacy and scientific literacy was not significant for both groups, as observed for the whole group model, and instrumental motivation to learn science remained the strongest predictor of scientific literacy (as a total effect) across both groups. Enjoyment of science also remained a strong predictor of scientific literacy through both direct and indirect effects. The relationships relating to epistemological beliefs about science were also similar for both groups, in terms of total, indirect and direct effects.

However, there are some notable differences between the male group and female group. This variation between groups is evident in terms of the significant difference in the model fit for each group as observed from the results of the chi-square difference test for the constrained models (regression coefficients held constant between the male and female groups) and the unconstrained models (regression coefficients were allowed to vary for each group) indicating that the differences in parameter estimates observed between the models were also of statistical significance.

Interest in broad science topics for the female-only group indicated a non-significant indirect effect (epistemological beliefs about science did not explain the relationship) and a *negative* relationship to scientific literacy scores (when controlling for ESCS and the other predictors in the model) leading to a net negative total effect ($B = -5.92$, 99% CI[-10.95, -0.90]). This negative direct effect was also observed in the male-only group, but the indirect effect was also significant and positive ($B =$ 3.90, 99% CI[1.73, 6.07]) resulting in the mitigation of the negative direct effect. In other words, the negative direct relationship between interest in broad science topics and scientific literacy (as discussed for the whole group model above) was offset by the positive indirect relationship between these two variables, through the improved sophistication of science epistemological beliefs. Again, this supports the notion that understanding the underlying processes for the development of scientific knowledge can bolster students' scientific literacy, and likely avoid the negative impacts of superficial interest, and thus understanding, of science that is likely susceptible to scientific misconceptions (Driver et al., 1985). Thus, for the male-only group, improved interest in science, to some extent, is related to improved epistemological beliefs about science and this positive relationship reduces the impact of the negative association between interest and scientific literacy in UAE government schools.

Another difference between groups was that of the total effect of instrumental motivation to learn science. For both groups, there was a small positive indirect effect transmitted to scientific literacy through epistemological beliefs about science (see Tables 2.4 and 2.5). However, the 5-point difference in scores between the male and female-only groups was due to a larger *direct* effect for males through some other mechanism than the development of their epistemological beliefs. There is some evidence that students in UAE public schools that express a higher likelihood to aspire to a career in science, technology, engineering and mathematical (STEM) fields perform considerably better in the scientific literacy assessment, and this effect is significantly stronger for male students than for female students (Cairns & Dickson, 2021). This strong relationship between aspirations and science performance for UAE males is likely the mechanism that results in an increased direct relationship between the instrumental motivation to learn science (where they respond to items that ask them how much they value their science learning in terms of their future career aspirations) and their scientific literacy test scores. However, it should be noted that over-reliance on extrinsic motivators (such as future job prospects) can lead to the development of poor study habits that reinforce negative attitudes and approaches to (lifelong) learning (Ames & Archer, 1988; Dweck, 2013; Meece et al., 1988).

Implications for Policy and Practice

The findings of this study provide strong evidence for the importance, for both males and females, of the development of students' epistemological beliefs about science in terms of student outcomes in scientific literacy. To support teachers, school leaders, and policymakers, recommendations for the development of epistemological beliefs in science classrooms are required. One such recommendation, to develop more advanced epistemic cognition, involves applying the AIR model (aims and value, epistemic ideals, and reliable processes), which involves three stages as implied by the name of the model (Chinn et al., 2014). Firstly, students can be encouraged to care about seeking the truth of a science concept by explaining its importance and *value* (aims and values stage). Students can also be supported in developing *ideals* relating to the importance of collecting reliable, bias-free scientific evidence (epistemic ideals stage). Following this, students can then develop a broad range of reliable processes for justifying scientific knowledge (reliable processes stage).

Additionally, students can be supported in moving towards an epistemic state referred to as "evaluativism". This state is hypothesised to be at the advanced end of a continuum of epistemological views that begin with *absolutists* (knowledge as absolute and certain), then moves to *multiplists* (sceptical of expertise and consider their views equal to experts, and gives more weight to emotions than evidence), and ends with *evaluativists* (Kuhn, 1991). Evaluativists tend to be less certain than experts and are able to evaluate conflicting viewpoints based on evidence. These views can be nurtured in the science classroom by stressing the importance of reliably gathered evidence and modelling how to evaluate competing claims about science that they have gleaned from their lessons and the wider community (Hofer, 2020).

It is also important to directly teach the epistemological assumptions of science. Science educators can explain the privilege of empirical research evidence over other methods of inquiry (Hofer, 2020). In this way, it is possible to challenge the misconceptions students may have relating to a scientific phenomenon by using the very processes of scientific knowledge creation to counter these beliefs (for example, students may not actually believe—despite being taught—that heavy objects fall at the same speed as similarly shaped, lighter objects without carrying out a systematic analysis of the phenomena through experimentation).

Other naïve beliefs about scientific knowledge may include the idea that science is a fixed body of knowledge (Hofer, 2004b). It is important to explicitly teach students about how scientific knowledge evolves through the generation and interpretation of new evidence. This further helps support the evaluativist empirical viewpoint as students are made aware that changes to scientific knowledge are not evidence that the experts are necessarily wrong but rather that the knowledge within the field has progressed (and will continue to do so).

Finally, another implication arising from this study is that UAE school students would also benefit from science interest-generating activities and experiences that are underpinned by clear learning goals relating to established processes of scientific knowledge development and justification. As such, "exciting" but ultimately superficial demonstrations of scientific phenomena (such as fun chemical reactions) that do not develop an understanding of the underpinning principles and processes of science appear to improve interest but lower science attainment for lower ability students. As such, demonstrations (and student-conducted experiments) should include elements of student activity that require them to explain the phenomena under study and state what evidence they have for their views by linking the observed phenomenon and the underlying science concept (Abrahams & Millar, 2008). Reflection activities that allow students to demonstrate their understanding of the reliability of the observed evidence in terms of the scientific inquiry procedures used to generate said evidence would also support their understanding of scientific epistemology and, ultimately, improve their scientific literacy.

Limitations of the Study and Directions for Future Research

The findings of this study should be considered alongside the limitations relating to the data and methods employed. Firstly, this study relies on self-reported survey data (not including performance in the scientific literacy assessment) relating to students' beliefs and attitudes to science learning. Self-reported data is subject to biases such as social desirability bias whereby the respondents respond to items in such a way as to conform to societal norms rather than express their true feelings (Grimm, 2010). However, the impact of these biases is minimised by the non-controversial nature of the items and constructs under study and the anonymity of the respondents in the publicly available datasets. Secondly, cultural issues relating to epistemological beliefs about science were not considered when interpreting the results of this study. The scale employed presents a Western worldview of epistemological beliefs about science coupled with the lack of acknowledgement of subcultures and indigenous beliefs outside of industrialised societies (Hofer, 2008).

Finally, the correlational nature of the study design means that it is only possible to discuss associations between variables and not causal relationships. The three main criteria for establishing strong evidence for causation (covariation between variables, the temporal distance between the hypothesised "cause" and "effect" variables, and the removal of alternative explanations) are only partially met in this study design (Hayes, 2017).

Future Research

Follow-up studies relating to this work should include the analysis of the epistemological beliefs scale employed by the OECD in the PISA 2015 assessment cycle in the context of UAE culture. The application, analysis and adaptation of the full four-dimensional epistemological beliefs about the science scale (on which the PISA scale was based; Conley et al., 2004) should be carried out in the UAE context. This work could involve validation processes such as cognitive interviews with members of the sample population, item response theory analysis, confirmatory factor analysis to verify the dimensionality of the instrument, and possibly adaptation of items to better reflect contextual cultural values. This analysis should also take place across government and private schools to take advantage of the diversity in the UAE school system to develop a more culturally inclusive measure for wider use. The development of such a measure could then lead to contextually appropriate and effective approaches to developing epistemological beliefs about science in the UAE education sector.

Conclusion

The findings of this study highlight the importance of students having clear understanding of the nature of scientific knowledge and the methods by which scientific knowledge is constructed. Epistemological beliefs about science are strongly associated with scientific literacy in UAE government schools both directly, and by mediating the relationship between the enjoyment of science and instrumental motivation to learn science. These beliefs also mitigate the negative relationship between interest in broad science topics and science outcomes for male students. The development of more sophisticated beliefs about science would support students' understanding of scientific phenomena and scientific investigation processes leading, ultimately, to improved outcomes for all students in the UAE government school system and beyond.

References

Abrahams, I., & Millar, R. (2008). Does practical work really work? A study of the effectiveness of practical work as a teaching and learning method in school science. *International Journal of Science Education, 30*(14), 1945–1969. https://doi.org/10.1080/09500690701749305.

Ames, C., & Archer, J. (1988). Achievement goals in the classroom: Students' learning strategies and motivation processes. *Journal of Educational Psychology, 80*(3), 260. https://doi.org/10.1037/0022-0663.80.3.260.

Areepattamannil, S., Freeman, J., & Klinger, D. (2011). Influence of motivation, self-beliefs, and instructional practices on science achievement of adolescents in Canada. *Social Psychology of Education, 14*(2), 233–259. https://doi.org/10.1007/s11218-010-9144-9.

Avvisati, F. (2020). The measure of socio-economic status in PISA: A review and some suggested improvements. *Large-Scale Assessments in Education, 8*(1), 8. https://doi.org/10.1186/s40536-020-00086-x.

Bandura, A. (1982). Self-efficacy mechanism in human agency. *American Psychologist, 37*(2), 122. https://doi.org/10.1037/0003-066X.37.2.122.

Bandura, A. (1997). *Self-efficacy: The exercise of control*. Macmillan.

Barron, K. E., & Harackiewicz, J. M. (2001). Achievement goals and optimal motivation: Testing multiple goal models. *Journal of Personality and Social Psychology, 80*(5), 706. https://doi.org/10.1037/0022-3514.80.5.706.

Bouffard-Bouchard, T., Parent, S., & Larivee, S. (1991). Influence of self-efficacy on self-regulation and performance among junior and senior high-school age students. *International Journal of Behavioral Development, 14*(2), 153–164. https://doi.org/10.1177/016502549101400203.

Britner, S. L., & Pajares, F. (2006). Sources of science self-efficacy beliefs of middle school students. *Journal of Research in Science Teaching: The Official Journal of the National Association for Research in Science Teaching, 43*(5), 485–499. https://doi.org/10.1002/tea.20131.

Browne, M. W., MacCallum, R. C., Kim, C.-T., Andersen, B. L., & Glaser, R. (2002). When fit indices and residuals are incompatible. *Psychological Methods, 7*(4), 403–421. https://doi.org/10.1037//1082-989x.7.4.403.

Brownlee, J., Boulton-Lewis, G., & Purdie, N. (2002). Core beliefs about knowing and peripheral beliefs about learning: Developing an holistic conceptualisation of epistemological beliefs. *Australian Journal of Educational and Developmental Psychology, 2*, 1–16. Retrieved from https://www.newcastle.edu.au/__data/assets/pdf_file/0015/100491/v2-brownlee-et-al.pdf.

Bryan, R. R., Glynn, S. M., & Kittleson, J. M. (2011). Motivation, achievement, and advanced placement intent of high school students learning science. *Science Education, 95*(6), 1049–1065. https://doi.org/10.1002/sce.20462.

Bryant, F. B., & Satorra, A. (2012). Principles and practice of scaled difference chi-square testing. *Structural Equation Modeling: A Multidisciplinary Journal, 19*(3), 372–398. https://doi.org/10.1080/10705511.2012.687671.

Burbules, N. C., & Linn, M. C. (1991). Science education and philosophy of science: Congruence or contradiction? *International Journal of Science Education, 13*(3), 227–241. https://doi.org/10.1080/0950069910130302.

Burns, E. C., Martin, A. J., Kennett, R. K., Pearson, J., & Munro-Smith, V. (2021). Optimizing science self-efficacy: A multilevel examination of the moderating effects of anxiety on the relationship between self-efficacy and achievement in science. *Contemporary Educational Psychology, 64*, 101937. https://doi.org/10.1016/j.cedpsych.2020.101937.

Byrne, C. M. (1996). *Water on tap: The use of virtual reality as an educational tool* (Publication Number 304270496) [Doctoral dissertation, University of Washington]. Available from ProQuest One Academic.

Cairns, D., & Areepattamannil, S. (2017). Exploring the relations of inquiry-based teaching to science achievement and dispositions in 54 countries. *Research in Science Education*, 1–23.https://doi.org/10.1007/s11165-017-9639-x.

Cairns, D., & Dickson, M. (2021). Exploring the relations of gender, science dispositions and science achievement on STEM career aspirations for adolescents in public schools in the UAE. *The Asia-Pacific Education Researcher, 30*(2), 153–165. https://doi.org/10.1007/s40299-020-00522-0.

Chen, J. A., & Usher, E. L. (2013). Profiles of the sources of science self-efficacy. *Learning and Individual Differences, 24*, 11–21. https://doi.org/10.1016/j.lindif.2012.11.002.

Chinn, C. A., Rinehart, R. W., & Buckland, L. A. (2014). Epistemic cognition and evaluating information: Applying the AIR model of epistemic cognition. In *Processing inaccurate information: Theoretical and applied perspectives from cognitive science and the educational sciences.* (pp. 425–453). Boston Review. https://doi.org/10.7551/mitpress/9737.003.0025..

Conley, A. M., Pintrich, P. R., Vekiri, I., & Harrison, D. (2004). Changes in epistemological beliefs in elementary science students. *Contemporary Educational Psychology, 29*(2), 186–204. https://doi.org/10.1016/j.cedpsych.2004.01.004.

Deci, E. L. (1971). Effects of externally mediated rewards on intrinsic motivation. *Journal of Personality and Social Psychology, 18*(1), 105. https://doi.org/10.1037/h0030644.

Driver, R., Guesne, E., & Tiberghien, A. (Eds.). (1985). *Children's Ideas in Science.* Open University Press.

Dweck, C. S. (2013). *Self-theories: Their role in motivation, personality, and development*. Psychology press.

Elder, A. D. (2002). Characterizing fifth grade students' epistemological beliefs in science. In B. K. Hofer & P. R. Pintrich (Eds.), *Personal epistemology: The psychology of beliefs about knowledge and knowing* (pp. 347–364). Routledge. https://doi.org/10.4324/9780203424964-26.

Fumerton, R. (2009). *Epistemology*. John Wiley & Sons.

Goldman, A. I., & McGrath, M. (2015). *Epistemology: A contemporary introduction*. Oxford University Press.

Gottfried, A. E. (1985). Academic intrinsic motivation in elementary and junior high school students. *Journal of Educational Psychology, 77*(6), 631. https://doi.org/10.1037/0022-0663.77.6.631.

Grimm, P. (2010). Social desirability bias. In *Wiley International Encyclopedia of Marketing*. https://doi.org/10.1002/9781444316568.wiem02057.

Harter, S. (1981). A new self-report scale of intrinsic versus extrinsic orientation in the classroom: Motivational and informational components. *Developmental Psychology, 17*(3), 300. https://doi.org/10.1037/0012-1649.17.3.300.

Hayes, A. F. (2017). *Introduction to mediation, moderation, and conditional process analysis: A regression-based approach*. Guilford publications.

Hofer, B. K. (2000). Dimensionality and disciplinary differences in personal epistemology. *Contemporary Educational Psychology, 25*(4), 378–405. https://doi.org/10.1006/ceps.1999.1026.

Hofer, B. K. (2004a). Epistemological understanding as a metacognitive process: Thinking aloud during online searching. *Educational Psychologist, 39*(1), 43–55. https://doi.org/10.1207/s15326985ep3901_5.

Hofer, B. K. (2004b). Exploring the dimensions of personal epistemology in differing classroom contexts: Student interpretations during the first year of college. *Contemporary Educational Psychology, 29*(2), 129–163. https://doi.org/10.1016/j.cedpsych.2004.01.002.

Hofer, B. K. (2008). Personal epistemology and culture. In M. Khine (Ed.), *Knowing, knowledge and beliefs* (pp. 3–22). Springer. https://doi.org/10.1007/978-1-4020-6596-5_1.

Hofer, B. K. (2020). Epistemic cognition: Why it matters for an educated citizenry and what instructors can do. *New Directions for Teaching and Learning, 2020*(164), 85–94. https://doi.org/10.1002/tl.20427.

Hofer, B. K., & Pintrich, P. R. (1997). The development of epistemological theories: Beliefs about knowledge and knowing and their relation to learning. *Review of Educational Research, 67*(1), 88–140. https://doi.org/10.3102/346543067001088.

Hu, L. t., & Bentler, P. M. (1999). Cutoff criteria for fit indexes in covariance structure analysis: Conventional criteria versus new alternatives. *Structural Equation Modeling: A Multidisciplinary Journal, 6*(1), 1–55.https://doi.org/10.1080/10705519909540118.

Jerrim, J., & Shure, N. (2016). *Achievement of 15-year-olds in northern Ireland: PISA 2015 national report*. UCL. Retrieved from https://www.education-ni.gov.uk/sites/default/files/publications/education/ACHIEVEMENT%20OF%2015%20YEAR-OLDS%20IN%20NORTHERN%20IRELAND%20PISA%202015%20NATIONAL%20REPORT..PDF.

Krapp, A., & Prenzel, M. (2011). Research on interest in science: Theories, methods, and findings. *International Journal of Science Education, 33*(1), 27–50. https://doi.org/10.1080/09500693.2010.518645.

Kuhn, D. (1991). *The skills of argument*. Cambridge University Press.

Lepper, M. R., Iyengar, S. S., & Corpus, J. H. (2005). Intrinsic and extrinsic motivational orientations in the classroom: Age differences and academic correlates. *Journal of Educational Psychology, 97*(2), 184–196. https://doi.org/10.1037/0022-0663.97.2.184.

Loehlin, J. C., & Beaujean, A. A. (2017). *Latent variable models: An introduction to factor, path, and structural equation analysis* (5th ed.). Routledge. https://doi.org/10.4324/9781315643199.

Mason, L., Boscolo, P., Tornatora, M., & Ronconi, L. (2013). Besides knowledge: A cross-sectional study on the relations between epistemic beliefs, achievement goals, self-beliefs, and achievement in science. *Instructional Science, 41*(1), 49–79. https://doi.org/10.1007/s11251-012-9210-0.

Mason, L., Gava, M., & Boldrin, A. (2008). On warm conceptual change: The interplay of text, epistemological beliefs, and topic interest. *Journal of Educational Psychology, 100*(2), 291. https://doi.org/10.1037/0022-0663.100.2.291.

Meece, J. L., Blumenfeld, P. C., & Hoyle, R. H. (1988). Students' goal orientations and cognitive engagement in classroom activities. *Journal of Educational Psychology, 80*(4), 514. https://doi.org/10.1037/0022-0663.80.4.514.

Muis, K. R., Pekrun, R., Sinatra, G. M., Azevedo, R., Trevors, G., Meier, E., & Heddy, B. C. (2015). The curious case of climate change: Testing a theoretical model of epistemic beliefs, epistemic emotions, and complex learning. *Learning and Instruction, 39*, 168–183. https://doi.org/10.1016/j.learninstruc.2015.06.003.

Muthén, L., & Muthén, B. (1998–2017). *Mplus user's guide* (8th, Ed.). Retrieved from https://www.statmodel.com/download/usersguide/MplusUserGuideVer_8.pdf.

Nugent, G., Barker, B., Welch, G., Grandgenett, N., Wu, C., & Nelson, C. (2015). A model of factors contributing to STEM learning and career orientation. *International Journal of Science Education, 37*(7), 1067–1088. https://doi.org/10.1080/09500693.2015.1017863.

OECD. (2016). PISA 2015 results (Volume I): Policies and practices for successful schools. *OECD Publishing*. https://doi.org/10.1787/9789264266490-en.

OECD. (2017). *PISA 2015 assessment and analytical framework*. https://doi.org/10.1787/9789264281820-en.

Ozkal, K., Tekkaya, C., Sungur, S., Cakiroglu, J., & Cakiroglu, E. (2010). Elementary students' scientific epistemological beliefs in relation to socio-economic status and gender. *Journal of Science Teacher Education, 21*(7), 873–885. https://doi.org/10.1007/s10972-009-9169-0.

Pajares, F., Britner, S. L., & Valiante, G. (2000). Relation between achievement goals and self-beliefs of middle school students in writing and science. *Contemporary Educational Psychology, 25*(4), 406–422. https://doi.org/10.1006/ceps.1999.1027.

Preacher, K. J., & Hayes, A. F. (2008). Asymptotic and resampling strategies for assessing and comparing indirect effects in multiple mediator models. *Behavior Research Methods, 40*(3), 879–891. https://doi.org/10.3758/BRM.40.3.879.

Qian, G., & Alvermann, D. (1995). Role of epistemological beliefs and learned helplessness in secondary school students' learning science concepts from text. *Journal of Educational Psychology, 87*(2), 282. https://doi.org/10.1037/0022-0663.87.2.282.

Rokeach, M. (1972). *Beliefs, attitudes and values: A theory of organization and change*. Jossey-Bass.

Schommer, M. (1993). Epistemological development and academic performance among secondary students. *Journal of Educational Psychology, 85*(3), 406. https://doi.org/10.1037/0022-0663.85.3.406.

Schommer, M., & Walker, K. (1995). Are epistemological beliefs similar across domains? *Journal of Educational Psychology, 87*(3), 424. https://doi.org/10.1037/0022-0663.87.3.424.

Schunk, D. H., Pintrich, P. R., & Meece, J. L. (2014). *Motivation in education: Theory, research, and applications*. Pearson Education Limted.

Sengul, O., Enderle, P. J., & Schwartz, R. S. (2020). Science teachers' use of argumentation instructional model: Linking PCK of argumentation, epistemological beliefs, and practice. *International Journal of Science Education, 42*(7), 1068–1086. https://doi.org/10.1080/09500693.2020.1748250.

Sinatra, G. M., Southerland, S. A., McConaughy, F., & Demastes, J. W. (2003). Intentions and beliefs in students' understanding and acceptance of biological evolution. *Journal of Research in Science Teaching, 40*(5), 510–528. https://doi.org/10.1002/tea.10087.

Singh, K., Granville, M., & Dika, S. (2002). Mathematics and science achievement: Effects of motivation, interest, and academic engagement. *The Journal of Educational Research, 95*(6), 323–332. https://doi.org/10.1080/00220670209596607.

Songer, N. B., & Linn, M. C. (1991). How do students' views of science influence knowledge integration? *Journal of Research in Science Teaching, 28*(9), 761–784. https://doi.org/10.1002/tea.3660280905.

Taylor, G., Jungert, T., Mageau, G. A., Schattke, K., Dedic, H., Rosenfield, S., & Koestner, R. (2014). A self-determination theory approach to predicting school achievement over time: The unique role of intrinsic motivation. *Contemporary Educational Psychology, 39*(4), 342–358. https://doi.org/10.1016/j.cedpsych.2014.08.002.

Thomson, S., & De Bortoli, L. (2008). Exploring Scientific Literacy: How Australia measures up. The PISA 2006 survey of students' scientific, reading and mathematical literacy skills. Retrieved from http://research.acer.edu.au/ozpisa/2/.

Trautwein, U., & Lüdtke, O. (2007). Predicting global and topic-specific certainty beliefs: Domain-specificity and the role of the academic environment. *British Journal of Educational Psychology, 77*(4), 907–934. https://doi.org/10.1348/000709906X169012.

Vallerand, R. J., Pelletier, L. G., Blais, M. R., Briere, N. M., Senecal, C., & Vallieres, E. F. (1992). The Academic Motivation Scale: A measure of intrinsic, extrinsic, and amotivation in education. *Educational and Psychological Measurement, 52*(4), 1003–1017. https://doi.org/10.1177/0013164492052004025.

Wenning, C. J. (2009). Scientific epistemology: How scientists know what they know. *Journal of Physics Teacher Education Online, 5*(2), 3–16. http://www2.phy.ilstu.edu/pte/publications/scientific_epistemology.pdf.

Wigfield, A., & Eccles, J. S. (2000). Expectancy–value theory of achievement motivation. *Contemporary Educational Psychology, 25*(1), 68–81. https://doi.org/10.1006/ceps.1999.1015.

Zhao, L., He, W., Liu, X., Tai, K.-H., & Hong, J.-C. (2021). Exploring the effects on fifth graders' concept achievement and scientific epistemological beliefs: Applying the Prediction-Observation-Explanation Inquiry-Based Learning Model in Science Education. *Journal of Baltic Science Education, 20*(4), 664–676. https://doi.org/10.33225/jbse/21.20.664.

Dr. Dean Cairns holds a PhD in Chemistry from the University of Sussex and is a doctoral candidate for an EdD in Education at Bath University. He has worked as both a research scientist and science educator. He is currently an Associate Professor in Curriculum and Instruction at the Emirates College for Advanced Education. He has published widely, in high-quality international journals in the fields of polymer chemistry and science education. Recent publications have included investigating the impact of inquiry-learning approaches on science outcomes, and students' perceptions of using technology during inquiry-related activities. His current research focuses on students' epistemic beliefs about science, and their relationships with science conceptual knowledge development.

Chapter 3
The Drawing a Scientist Test (DAST): How Do Girls in the UAE Present Visual Characteristics of Female Scientists, and What Does This Mean for Gender Equity of Science Careers?

Martina Dickson and Melissa McMinn

Abstract The 'Drawing A Scientist Test' (DAST) has been used by educators for decades for its ability to act as a 'window' into children's perceptions of scientists and the work that they do. Although the method is not without its critics, researchers have tentatively extrapolated the analysis of features in such drawings to infer students' science aspirations. This study takes place in the United Arab Emirates (UAE), where the national economy is moving towards a predominantly STEM-based knowledge economy, which requires high levels of workforce participation. In this chapter, we focus on the ways in which female students ($n = 107$) graphically depicted scientists in a DAST. The majority of girls drew female scientists, and a significant minority of these female scientists (around 20%) exhibited symbolism of hyper-femininity through garments, accessories, hair style, and make-up. The potential implications of these perceptions to girls' science identities and aspirations are discussed.

Keywords Drawings · Identities · Perceptions · Scientists · Work

Introduction

This study takes place in Abu Dhabi, the capital and largest emirate of the United Arab Emirates (UAE). This is a country that has experienced rapid development of its society and economy since the discovery of oil in the late 1950s. Over the last two decades, in particular, vast strides have been made in educational reform, through various high profile, high investment government-sponsored initiatives. STEM initiatives and goals are heavily prioritised in the National Vision 2030 (Abu Dhabi

M. Dickson (✉)
Emirates College for Advanced Education, Abu Dhabi, UAE
e-mail: martina_dickson@hotmail.com

M. McMinn
Open Polytechnic, Lower Hutt, New Zealand

M. Dickson et al. (eds.), *Gender in STEM Education in the Arab Gulf Countries*,
https://doi.org/10.1007/978-981-19-9135-6_3

Economic Vision Report 2030, 2008), which requires science and technology graduates from universities to fulfil the country's ambitions of building a knowledge economy. Having science graduates requires aspiring students to select those degrees in the first place, and girls and women in STEM have received much focus from the UAE's leadership. Therefore, an examination of students' perceptions of scientists and the work they do is important, due to the previously researched and established relationship between positive perceptions of scientists and uptake of science degrees and careers.

In this chapter, we build upon existing research and use school students' drawings of scientists to identify characteristics that they associate with scientists and their work, with a view to extrapolating these to possible future science career resonance. We focus particularly on the characteristics of scientists that female students draw, since research has suggested that girls may hold preconceived gendered stereotypical notions of science and scientists, such as science not being 'feminine' enough to appeal to girls for whom feminisation is important. We aimed to explore these ideas, and to find out how 'feminised' the characteristics of drawings of scientists were.

Literature Review

Drawing a Scientist Test

Graphical depictions of scientists have been used as a 'window' into children's thoughts and perceptions of scientists since the 1970s. In 1983, David Chambers developed the DAST (Drawing A Scientist Test) protocol, and analysed over four thousand drawings for elements such as gender, presence of laboratory equipment, safety glasses, lab coats, symbols of danger, etc. (Chambers, 1983). Finson et al. (1995) developed a checklist corresponding to those, and other, characteristics to facilitate the analysis of the drawings. Later modifications of the study also included the identification of mythical characteristics, such as similarities to Frankenstein's monster.

Since then, the study has been replicated internationally, for example in the U.S.A. (Farland-Smith, 2012; Finson et al., 1995; Mason et al., 1991), Turkey (e.g. Akcay, 2011; Ozel, 2012; Türkmen, 2008), Australia (Milford & Tippett, 2013), and Europe (e.g. Christidou, 2010; Reinisch et al., 2017). Some patterns of drawing features have remained consistent over time. In most DAST studies, scientists are drawn wearing lab coats and safety glasses, and are invariably performing chemistry work. This work often involves danger; classified by characteristics such as warning signs for poison, toxic gases, etc., with some variation of the extent of these characteristics (e.g. Christidou et al., 2012; Fung, 2002; Ozel, 2012; Türkmen, 2008).

One characteristic which has shown slow growth over time and geographical region, though, is the portrayal of gender. When we carried out our own DAST during the autumn of 2019 (Dickson & McMinn, 2022), we were interested to find that many patterns did echo those of other studies, such as high frequency of reference to the symbolism of danger and the propensity of chemistry experiments, with test-tubes and bubbling flasks galore. Our study did differ from others in two important ways. Firstly, it takes place in the Gulf region of the Middle East, and to our knowledge this was the first time such a study had been undertaken in the region, and so it was unique in that sense. Secondly, most children drew a scientist of their own gender, and a much higher percentage of girls drawing female scientists (79%) was observed than has been noted in other studies. This was exciting and significant for a number of reasons. Drawing on theoretical concepts of depictions acting as projections of ideas about who scientists might be, the drawings we collected from both girls and boys suggested that perhaps being a scientist *was* a possibility for them, but also in that the representation of one's own gender is highly suggestive that those careers are possibilities for their gender, removing one barrier at least.

In the UAE, as in many countries in of the Arab world, the gender gap in STEM achievement highlighted by large-scale secondary dataset analyses is seen in the reverse of that in other world regions. Girls frequently outperform their male counterparts in science, maths, reading and writing, in UAE government schools where the majority of students are national citizens (OECD, 2014). In the UAE, higher education institutions receive greater numbers of female than male applicants and enrollees in physics degrees, for example.[1] The educational development of girls and women in the UAE has been rapid, and from the 1960s to the present day, women have been encouraged to be part of the STEM workforce in both policy and practice.

In the DAST studies we have previously referred to, researchers have replicated the original features as laid out by the Chambers (1983) and Finson et al. (1995) studies, with little variation. This was for good reason, such as wishing to replicate for comparison with other studies and to track attitudes changing over time within a region. The standard characteristics analysed did not, however, provide in-depth insight into the gendered features of the scientists. We wished to take a closer look at the images of scientists that girls were drawing, beyond standard DAST characteristics.

Theoretical Framework

> *Performing gender is not straightforward; rather, it is confusing.*
>
> (Reay, 2001, p. 163)

We wondered, since a higher % of girls in our study drew female scientists than is often observed in other studies, *how* did they use drawings to construct their imaginary female scientists? What particular characteristics of female scientists did they depict, and what indications of social constructions of the role might influence these characteristics? The underlying premise (and presumably, popularity) of the

[1] Personal communication with University Departmental Chairs.

DAST is that the drawings may offer a 'window' into perceptions of what scientists do, and critically—*who* scientists are. Girls drawing female scientists implies two things; one is the assumption of sociocultural norms ('women can be scientists'), and the other is a possibility for self ('I, as a girl, can one day be a woman scientist, if I choose to'). It is sometimes assumed that students are more likely to aspire to study science and choose science career paths where they are exposed to role models with whom they could possibly identify. These could be teachers of the same gender, ethnicity, community or anyone held in high status positions whom children are likely to look up to. However, Schinske et al. (2016) theorise about 'possible selves' as a more meaningful framework for self-identification than role modelling. Instead, the 'possible self' is a model whereby students begin to see themselves as scientists, with their own work in science class as part of that growing identification; a subtle yet important difference to the role model. In the context of drawings, girls may be identifying with their 'possible selves' as scientists, particularly where the control 'person' drawing also resembles themselves.

In the 1950s, pioneering sociologists Mead and Metraux (1957), articulated the complexity (and danger) of directly correlating imagery with students' personal choices for themselves: "the official image of the scientist—that is, an image that is the correct answer to give when the student is asked to speak without personal career involvement … that is not so when the student's personal choices are involved" (p. 384). We also deliberately withhold from using our study findings to directly correlate with career projections, yet we do believe that the characteristics of the drawings which are depicted may well represent underlying perceptions of what scientists do, and importantly *who* scientists are.

Archer et al. (2013) describe girls' constructions of 'desirable femininity' within career aspirations contexts and showed that girls often aspired to careers that were one of two main themes: 'nurturing' (e.g. aspiring to careers such as teaching, nursing, etc.), or 'glamorous/girly' with a "stereotypically feminine flavour to girls' aspirations for 'glamorous' careers (notably in acting, dancing and singing)" (p. 179). Girls who socially constructed desirable femininity in these terms referenced science careers against these norms, such as expressing views that science was not nurturing enough, or that it was unlikely (presumably) to provide an outlet for desirability for glamour or 'girliness'. As Archer et al. (2013) put it: "science did not seem to be popularly perceived as congruent with performances of ('girly') popular hetero-femininity" (p. 180). The notion of caring, nurturing professions (which they did not identify with science careers) also corresponds with conformity to being 'good girls' (Francis et al., 2017a). This notion may persist even in contexts where girls outperform boys academically (as in the UAE), and this recognition of competing gender discourses is important since some discourses are more powerful than others for some girls. Girls' constructions of themselves and others as being a powerful force in identity, particularly the 'girly/glamorous' ideas, is a concept echoed by Read (2011), who describes how gendered social constructions of popularity (termed 'the popular girl' in the paper) mix with girls' ideas of their own role models, as well as future career aspirations. Similarly to the Archer et al. (2013) study, Read's (2011) study described young girls aspiring to 'girly/glamorous' dominant feminine constructions (the female singer and performer Beyonce was named as a favourite): "for many girls in our study the constructed images of such female celebrities represented more than anything else their own future ideal self-identity—who they would like to be when they grow up" (p. 4).

In a study of teenagers' potential uptake of, and attitudes towards physics, the assertion that 'girly-girls' were less likely to want to pursue the subject was often attributed to social stereotypes associated with femininity, such as a lack of the required knowledge, concentration, or independence, and an aversion to manual and/or dirty work—by both female and male students (Francis et al., 2017b). In this study, 'girly-girls' were also considered by some students to be heavily influenced by their peers, and science was not within the realm of socially acceptable pursuits for this group.

One issue with these types of constructed aspirational characteristics are that those desired characteristics can "be seen to emphasize passivity rather than agency and power … [with] emphasis on attractiveness and appearance rather than activity and accomplishments" (Read, 2011, p. 1). Given that concepts of femininity and masculinity are essential concepts of relation, (Francis et al., 2017a) that is, they are true only in relation to the other, characterising science as not being feminine for any reason, including it not being glamorous, implies that science is therefore masculine. These concepts of masculinity or femininity are not fixed in performance but change slowly in social processes which are ongoing (Orlander, 2019). Where girls project constructions of feminine characteristics as desirable, science may not be considered 'girly', therefore science may be non-desirable to girls seeking popularity, or in those thinking ahead to possible future selves. Science becomes, therefore, an "imagined space that is incompatible with girls' performances of popular/desirable hetero-femininity" (Archer et al., 2013, p. 181). Archer et al. explain this as therefore indicating science career constructions as being 'masculine', by contrast.

Betz and Sekaquaptewa (2012) found that, contrary to popular belief, feminine STEM role models may in fact discourage girls who already do not identify with STEM, possibly because it confuses their constructions of what it means to be feminine, which do not match what it means to be a scientist. Francis et al. (2017a) also showed that this "affirmation that girly/super-feminine girls were less likely to want to pursue physics was often ascribed to social stereotypes associated with femininity" (p. 1103). Role models such as highly successful women who appear to be so far away from someone's possible self or experience, so much so that they appear 'unobtainable', can also make people feel discouraged and threatened, ultimately resulting in a belief that such a level of success was unobtainable (Latu & Mast, 2015).

Authors' Positionalities

Both authors are teacher educators and were previously science classroom teachers for children of primary, middle or high school ages. Our researcher positionalities tend to be based on feminist theoretical stances, and our social identities as both experienced schoolteachers and parents of school-age children ourselves has definitely influenced our interest in, and choice of, this particular field of research. We have discussed at length with one another, ways in which these positionalities and identities could possibly influence assumptions we have made about the images drawn by the students, or interpretations of the data. Accordingly, from the outset of the project we established a rigorous methodology for the interpretation of the students' drawings in order to retain objectivity in our interpretation. It obviously still influences our discussion of the findings, but by presenting the findings informally at our workplaces and the resulting conversations with peers, we felt that we had presented a discussion of the findings which was fair and balanced.

Methodology

The participants were students aged between 8 and 12 years old, (between grades 3 and 6), and drawn from public and private schools in the UAE's capital, Abu Dhabi. Upon completion of the processes to receive the relevant research approval from our institutions and the local educational administrative authority, we communicated with school administrators in order to seek permission to carry out the study with consenting students. We also sought the school administrators' assistance with the dispensing of the blank, coded, drawing papers. The packs which we distributed contained these papers, coloured pencils, and included instructions for teachers to read out when introducing the task to students. These instructions were stressed so that verbal directions given to students would be consistent across classes and schools, to avoid teachers potentially influencing students' drawings. Many critiques of the DAST have centred upon methodological issues such as lack of consistent instructions to students, not providing coloured pencils (therefore limiting how students can present drawing elements including skin tones), not having a controlled drawing of a 'person' or other subject (e.g. Hill & Wheeler, 1991; Losh et al., 2008; Reinisch et al., 2017). We were keen to ensure our methodology took those criticisms into consideration and avoided such issues.

Drawings were collected from 107 female students and 127 male students (total $n = 234$). We present the findings of the gender of the scientists that the male students drew, for a holistic picture of the context, but our analysis and discussion focus on the girls' drawings. The protocol for administering the test in the classroom was simply that the regular classroom teacher asks the students to "draw a picture of the scientist doing science" without any previous discussion, as per the original study instructions. They were also asked to 'draw a person' as per the standard protocol of other (including the original) DAST studies. The instructions were written in both English and Arabic, in order to support students who had Arabic as their first language. The time for complete administration of the drawing task was projected to be around 20 min.

Each drawing was rated (1-present, 0-not present) for the presence of specific characteristics. We checked for inter-coder reliability by each researcher coding a 'blind' sample of around 20 drawings and comparing researcher responses. Samples were coded iteratively until no inter-coder variability was observed. The drawing characteristics which were analysed were the same as those in the original DAST, for example eyeglasses, wearing lab coats, scientific instruments or laboratory equipment, symbols of knowledge (e.g. books), symbols of technology (the 'product of science), captions such as formulae (Chambers, 1983).

We also noted characteristics quantified by researchers such as Finson et al. (1995), and Barman (1997) who had rated incidences of mythical stereotypes such as the monster which Frankenstein created, as did we. Finally, since we were interested in the characteristics of female scientists which were drawn by girls, we paid close attention to the physical characteristics of these depicted women, particularly features which could denote symbolism of femininity or masculinity such as hairstyles, hair accessories, colours of clothing, types of clothing, presence of jewellery, style of footwear, make-up and motifs of symbols on clothing. We were also alert to potential indicators which indicate feminine cultural nuances of the UAE, which we planned to quantify should such data arise.

Research Ethics

In order to ensure the protection of the vulnerable population—the students—involved as participants in this research study, we first applied to our IRB with details of how we would seek student assent to partake in the drawing activity. We provided guidance notes for the teachers about alternative activities should the student not wish to partake in the activity. We also provided clear written instructions for the teachers administering the drawing instrument, including that the students should not put their names on the drawings. This measure ensured that the drawing remained anonymous, and the only data available to us was the self-identified gender of the child (identifiable by the colour of the drawing paper only).

Findings

Of the drawings collected, 79% of girls drew female scientists, compared to 81% of boys who drew male scientists (see Table 3.1). 19% of girls drew male scientists, while the remaining 2% drew scientists which we as researchers both concluded independently to have 'non-identifiable gender'. In contrast, only 2.4% of the boys drew female scientists. A higher proportion of females are represented in the girls' drawings of people (92%), compared to the girls' drawings of scientists. Girls' drawings of female scientists depict, on the whole, fairly similar characteristics to those observed in other studies, with lab coats featuring frequently (70% of drawings), and just over a third (35%) of the scientists were drawn wearing safety glasses. The vast majority of scientists were using lab equipment, typically glassware such as flasks, test tubes, beakers, etc., and typically conducting chemistry experiments. Almost half (44%) of the girls' drawings showed scientists doing some kind of dangerous work, either through symbolism or slogans of the 'Danger! Keep out!', nature, or through bubbling, toxic-looking gases.

We look now in more detail at the characteristics of the female scientists which were drawn by girls, paying close attention to instances of hyper-feminised symbols; symbols that deliberately and directly demonstrate the student's alignment with the construction of 'feminine' markers. Of the 79% (85) of female scientist drawings produced by girls, around half depicted the scientist with hair pulled back, (21% in high ponytail styles, 10% with hair bun styles, 20% in other pulled back styles). In the drawings of people, this figure was much less (as would be expected in a non-work environment) with more female people seen with loose flowing hair. Where students

Table 3.1 Comparing girls' and boys' depictions of the scientist's tools and working environment

Drawing characteristic		Girls' drawings (%) (*n* = *107*)	Boys' drawings (%) (*n* = *127*)
Drew a female scientist		79	2
Drew a male scientist		19	81
Drew a female person		92	2
Drew a male person		4	80
		Girls' drawings of female scientists (%)	
Scientist characteristics	Lab coat	70	
	Safety eyeglasses	35	
	Lab equipment	89	
	Knowledge symbols	18	
	Technology	3	
	Danger or fear symbolism	44	
	Captions/Formulae	9	
	Mythical features	2	

chose to colour their figure's hair, we recorded 32% of the scientists with dark hair, and 12% with light coloured hair (penciled in yellow, orange or similar), which was marginally similar to the hair colours in the drawings of people. The presence of accessories in the scientists' hair was not uncommon; this was observed in 20% of the drawings and were items such as hairbands, hair bows, hair clips, frequently pink in colour. These were observed much more often in the scientists' drawings, compared to the drawings of people, in line with the hair being more frequently pinned up. A quarter of the female scientists were drawn wearing emphasised make-up (mascara, lip colour, and sometimes cheek blush colour). We recorded these items only where we could say emphatically that the colour had been emphasised by the student. Noticeable long eyelashes were depicted in 32 of the people drawings (35%), and 20 of the scientist drawings (25%). The presence of jewellery was limited, and only one scientist was seen wearing a jewellery item (a pearl necklace). This figure was slightly higher in the drawings of people, where six items of jewellery were noted (necklaces and bracelets). We also observed four female scientists wearing knee-high boots (e.g. Fig. 3.1), although this was also observed in the peoples' drawings with the same frequency. We observed six instances of high-heeled shoes in the drawings of people, and none in the drawings of scientists.

The drawings were sometimes observed to depict the same person in both drawings, usually with key additional features in the scientist drawing such as the lab coat, safety glasses, and usually wearing more conservative clothing (see, for example, Fig. 3.2a). Less pink items of clothing are observed in the scientist drawings, less skirts, and more sensible shoes (as one would expect) in the scientist drawings than in the drawings of the person. Other drawings demonstrated a stark contrast between the person and the scientist, for example, clear hyper-feminised symbols in the 'person', much more so than in the scientist. Figure 3.2b shows an example of this. Note that the hair of the person is shorter, and they are wearing a lab coat and safety glasses (albeit on top of her head), but with the token pink shirt underneath to denote femininity, compared to the corresponding person drawn by the same student, with extremely long hair, and the off-the-shoulder style short dress. This contrast was often emphasised when the student had drawn a female person and a male scientist. See Fig. 3.2c for one example of this, where the characteristics of the female person are particularly feminised in direct contrast to the male scientist: pink hairband, pink jersey with love hearts, necklace and make-up.

The drawings of people showed clothing that was, on the whole, much more colourful, dramatic and feminised than the scientist: some examples of clothing observed in these drawings included pink tops and dresses, sometimes adorned with love hearts, shirts with 'fun' motifs (e.g. "hi!") and symbols (suns, flowers, stars), long yellow, purple and pink dresses, patterned items such as animal print and polka dot skirts, some instances of off-the-shoulder type tops, and some ballerina outfits.

Fig. 3.1 Drawings of a female scientist wearing knee-high boots and jeans

Discussion and Implications

Girls Drawing Female Scientists

The numbers of drawings of female scientists drawn by girls in this study indicate that they consider that being a scientist is a role for their gender. More than three-quarters of the girls drew female scientists, and while this remains a smaller proportion than the boys who drew male scientists, this is a significant increase in comparison with

(a)

Fig. 3.2 Drawings of a scientist and person by the same female student

many previous DAST studies. Barely over 1% of the girls in the original DAST study (Chambers, 1983) drew female scientists. Since that time, most studies have recorded the total number of female scientist drawings (always a minority), without identifying the gender of the drawers, however, several studies (see, for example, Bodzin & Gehringer, 2001; Buldu, 2006; Rosenthal, 1993), have pointed out that almost all female scientists were drawn by female students. A similar percentage of

(b)

c)

Fig. 3.2 (continued)

girls drawing female scientists as in the current study was recorded in a Greek study with similar-aged students (Emvalotis et al., 2018).

Girls Seeing Science as a Career for Themselves

It is less clear how the drawings of female scientists relate to the girls' projections of their possible selves as scientists. For those 20% or so girls who have drawn feminised scientists, do the symbols of glamour and hyper-femininity indicate a new stereotype of the female scientist, one which conforms to stereotypical views of girls as needing to be 'girly', 'feminine', etc.? Or do these hyper-feminine drawings indicate a scientist which they believe exists, but not a stereotype to which they themselves aspire? Where students appear to have drawn the same character as the 'person' and the 'scientist', this may be a reflection of their own self, suggesting they view science as a viable future career for themselves. Rather than perceive scientists as being in a profession which is not feminised, or that scientists themselves are not 'feminine' enough, these new feminised images of scientists may indicate a liberation of that previous stereotype, and perhaps this possibility ('scientists can also be feminine!') appeals to those girls for whom being 'girly/glamorous' is important, and who would otherwise self-select away from science.

Stereotypes

Girls depicting scientists as these hyper-feminine women are not without issue. They do appear contradictory to the work of others, for whom being girly, 'glam', etc., was synonymous with not being 'like a scientist' (Francis et al., 2017b). But for girls who do not espouse—or do not wish to espouse—such characteristics, stereotypes of glamorous female scientists with high swinging ponytails, knee-high boots and pink love-heart emboldened tops, may also be prohibitory. Indeed, as mentioned earlier, there is some evidence that having highly feminised role models in STEM can decrease the interests of girls who have previously articulated being interested in STEM (Betz & Sekaquaptewa, 2012) due to them being perceived as unobtainable roles by the girls, thus demotivating rather than motivating. We would argue that it is just as undesirable to have a situation where scientists are imagined as hyper-feminine figures, than to have all female scientists drawn with a serious expression, thick glasses and brown hair. The stereotype itself, and by definition its lack of diversity, is the barrier. It may be positive, then, that the female scientists drawn in the current study reflect features of hyper-feminisation in only a proportion of drawings. While these features are less evident in drawings of scientists than in drawings of people, this may simply reflect the work environment of the scientist (as opposed to drawings of the person, who was not depicted in such an environment). It suggests that some students believe that women can be scientists, involved in often dangerous work, without compromising their femininity. This is in contrast to the Francis et al. (2017b) study, in which the student participants shared views of the characteristics of those who do physics and characteristics of 'girly-girls' to be mutually exclusive. Even with the lab coat left on, these drawings would have been equally appropriate in a beauty salon or nursing setting (careers deemed more appropriate for 'girly-girls' by the students in the Francis study, 2017b) as in the chemistry lab, suggesting more diversity in the potential careers for these 'girly-girls'.

However, the current study also produced drawings of female scientists who wear clothing in types and colours less associated with femininity, and without make-up, jewellery, and other accessories—suggesting that a strong stereotype of a female scientist, overall, was not omnipresent.

Work of the Scientists

The work of the female scientists depicted in the girls' drawings largely reflected that seen in previous DAST studies. The vast majority of female scientists were drawn working alone in laboratory settings, using equipment related to chemistry experiments, often with lab coats on. This may indicate a narrow view of the work which scientists do, which in turn could be prohibitory for girls when considering science as a career. Only one drawing by a female student was a scientist working outside, near mountains, with the caption 'examining minerals', however, this scientist was drawn from behind, and therefore the gender was unidentifiable. All of these scientists were portrayed as working alone. This is in line with the results of several previous studies (see, for example, Buldu, 2006; Chambers, 1983; Emvalotis & Koutsianou, 2018; Hillman et al., 2014; Turkmen, 2008), and intimates a strong and persisting perception that science is an individual endeavour. It has been suggested that this could, in part, be explained by the instruction to draw a 'scientist' (singular) but may also be considered a deterrent by young girls to whom socialising and friendship circles are often very important.

It was interesting to note that in the current study, almost half (44%) of the girls' drawings had the symbolism of fear or danger (Table 3.1). In previous studies, danger featured in frequency in only a fifth of the girls' drawings when compared to boys' drawings (Emvalotis & Koutsianou, 2018). However, in the current study, it was the girls who drew slightly more images of fear or danger, which may be attributable to the prevalence of experimental chemistry work shown in the drawings. It is not clear whether the element of danger is something that appeals to girls, or whether it is seen as a deterrent. Noticeably missing from the girls' drawings were symbols of technology, perhaps surprising given the prevalence of technology in the girls' everyday lives. Again, this could suggest a narrow definition of the work scientists do, and the absence of technology may or may not be appealing to girls contemplating science.

Girls Drawing Male Scientists

If we assume that girls drawing female scientists suggest they see this career as one for their gender, we could equally assume that in the cases (19%) where girls drew male scientists, they may not hold this view. Again, this would need further investigation to corroborate, however, in all cases where girls drew male scientists, they drew a corresponding female person, which suggests a greater distance between 'possible selves' and a career in science.

Science Capital

We were also interested in considering the drawings through the lens of the concept of 'science capital', defined by Archer et al. (2014, p. 5) as:

> A conceptual device for collating various types of economic, social and cultural capital that specifically relate to science - notably those which have the potential to generate use or exchange value for individuals or groups to support and enhance their attainment, engagement and/or participation in science.

Archer et al. (2014) argued that scientific forms of cultural and social capital could command a high symbolic and exchange value. Given the propensity of glamorous scientist drawings with these feminine symbolisms, we wonder if these characteristics are, in fact, also forms of science capital. We suggest that these may be potentially expanding the science capital jurisdiction.

Research from DeWitt et al. (2016) indicated that science capital was closely related to cultural capital, and that "particular dimensions of science capital (science literacy, perceived transferability and utility of science, family influences) seem to be more closely related to anticipated future participation and identity in science than others" (p. 2431). This deep and reflective consideration of participation is important internationally, but particularly in countries like the United Arab Emirates which are educationally developing, and for whom the participation of citizens in its STEM workforce is such an integral part of a long-term economic sustainability plan. The key social justice element of science capital, and the way in which it potentially serves social justice and equity for both boys and girls in science participation, means that science capital can serve as a unique conceptual lens to help in the understanding of uneven science participation patterns (Archer et al., 2012) in order to identify patterns in the UAE, and posit implications for schools and in particular, for the science classroom.

Young people with low levels of science capital may not identify themselves with science and therefore may be less likely to aspire to science careers. Since teachers can have a significant difference in how students engage with science, they can really capitalise on the experiences and interests which students have, potentially improving their engagement by doing so (Godec et al., 2017).

Importantly for this study, science capital is also seen to strongly interact with gender and science identity. In Archer et al.'s (2014) narrative study of teenage boys, beliefs were presented about the 'types' of people who could study science, use science and *be* scientists—such as 'brainy' and middle class (as opposed to the working class) boys. Other studies explain large under representation of females in particular fields of science degree (namely, physics and computer science), fields where "practitioners believe that raw, innate talent is the main requirement for success, because women are stereotyped as not possessing such talent" (Leslie et al., 2015, p. 262). Such women would not be seen to be possessing science capital. Leslie et al. also explain the stereotyping of women by others, as "lacking innate intellectual talent" (p. 264) and so, science capital. These notions of competence are an important element of examinations of the concept of science capital and may be connected in the ways in which the girls depict themselves, and the drawings of female scientists. Archer et al. (2015) cite examples such as girls with low science capital exhibiting low science self-efficacy, and poor science identity—not seeing themselves, and not thinking that other people saw them—as being science 'people'. Bringing together the key findings of research groups such as Leslie et al. (2015), and Archer et al. (2015), these gendered indicators of science identity and science capital, are indicative of the likelihood to participate in science in both school and beyond. Again, this is particularly important in countries such as the UAE for aforementioned reasons. The key social justice element of science capital, and the way in which it potentially serves social justice and equity for both boys and girls in science participation, means that science capital can serve as a unique conceptual lens to help in the understanding of science participation patterns.

In the UAE context, a country in which both authors have both worked and researched in educational settings for more than a decade, no such analysis has been performed on the science capital of the students in the UAE. How powerful are constructs of science capital in a country with a developing education system, where both genders' performance in science consistently falls below the OECD average (OECD, 2019), and where boys are consistently seen to be outperformed by their female counterparts in both science attainment and participation in post-secondary education? Is it possible that some of the features of the girls' drawings, where it might appear that those scientists depict science capital indicators, suggest that girls who drew the 'person' and the 'scientist' as the same person, possibly though not conclusively themselves, perceive scientists to indeed, possess science capital? This is an interesting area for future research pursuit.

The Influence of Media

Some previous DAST studies have suggested the media are an important source of influence on students' perceptions of scientists (Buldu, 2006; Song & Kim, 1999). The findings from this research have several implications for the design of STEM intervention programmes for girls. Many such intervention programmes for girls use media resources such as educational videos, computer games, radio programmes, and television programmes—in efforts to influence girls' perceptions of scientists. These programmes should acknowledge the potential influence of cultural images of gender, consider the use of media images of engineers and scientists in interventions focused on stereotypes and counter stereotypes, and assess girls' responses to specific stereotyped and counter stereotyped characteristics found in media images of women scientists. Others have found that images of scientists are less likely to be influenced by media than science teachers and images from textbooks (Turkmen, 2008) suggesting that these also should include a variety of female scientists to dispel predominant stereotypes.

Study Limitations

Our study, as with any, has limitations. For example, we did not record the gender of the participants' teachers in this study. Given that teachers have been found to influence students' perceptions of scientists (Turkmen, 2008), it may have been interesting to note this, to see whether such influence exists. In future studies examining girls' perceptions of scientists, it is recommended that the gender of science teachers be recorded and explored for correlation. This study included participants from five different schools in Abu Dhabi. The characteristics of these schools indicated that they could be considered to be representative of students across the Emirate, but are still fairly limited in number. An expanded study throughout the UAE would allow for a greater exploration of girls' drawings of scientists. For future work, we plan to undertake further research where we could ask students to draw themselves as the control person, to undertake those comparisons of self-identity with scientists to observe the likelihood of scientists as 'possible selves'. We also plan to draw on the recommendations of others (e.g. Akcay, 2011; Hillman et al., 2014) and complement the drawings with interviews to explore the thoughts behind the drawings, and whether certain features of their scientist and the work they are drawn doing are appealing, and to triangulate this with children's science career aspirations.

Conclusion

With a heavy priority on STEM initiatives and goals (as laid out in the National Vision 2030), the UAE requires science and technology graduates who can fulfil the country's ambitions of growing the knowledge economy. Girls and women in STEM have received much focus from the UAE's leadership, but little data exists regarding the aspirations of this key demographic to enter into STEM-related fields, such as the sciences. Results of the test presented in this article show that many stereotypical features of scientists have persisted over time, such as the propensity for chemistry work, lab coats and safety goggles. One significant change in the current study is the proportion of female scientists drawn, with 73% of scientist drawings depicting females. This suggests, positively, that science is perceived by the participants in this study as a career for women, although it is not known whether the girls in the study wish to pursue such a career themselves. The results showed that representations of hyper-femininity were found in approximately 20% of the female scientist drawings in the form of clothing, accessories, make-up, and other feminine symbolism. This indicates that there is science resonance for those girls identifying as being girly/feminine, but also that the majority of girls did not draw hyper-feminised female scientists. As there was a variety of styles of dress and appearances of the female scientist drawings rather than evidence of conformity to one stereotypical image (as is frequently seen in boys' drawings of scientists—the 'mad scientist'), this may suggest the dissolving of a predominant stereotype of female scientists. Given that stereotypes can act as a barrier to students pursuing various career paths, this is seen as an encouraging sign. Further research is planned to expand the current study in the UAE, and to follow up the DAST tests with interviews to allow students to explain their drawings (and underlying influences) and discuss their science career aspirations to allow for correlation with their drawings.

References

Abu Dhabi Economic Vision Report 2030. (2008). The Government of Abu Dhabi. Retrieved from https://www.ecouncil.ae/PublicationsEn/economic-vision-2030-full-versionEn.pdf.

Akcay, B. (2011). Turkish elementary and secondary students' views about science and scientist. *Asia-Pacific Forum on Science Learning and Teaching, 12*(1), 1–11.

Archer, L., DeWitt, J., Osborne, J., Dillon, J., Willis, B., & Wong, B. (2012). Science aspirations, capital, and family habitus: How families shape children's engagement and identification with science. *American Educational Research Journal, 49*(5), 881–908. https://doi.org/10.3102/0002831211433290

Archer, L., DeWitt, J., Osborne, J., Dillon, J., Willis, B., & Wong, B. (2013). 'Not girly, not sexy, not glamorous': Primary school girls' and parents' constructions of science aspirations. *Pedagogy, Culture & Society, 21*(1), 171–194. https://doi.org/10.1080/14681366.2012.748676

Archer, L., Dewitt, J., & Willis, B. (2014). Adolescent boys' science aspirations: Masculinity, capital and power. *Journal of Research in Science Teaching, 51*(1), 1–30. https://doi.org/10.1002/tea.21122

Archer, L., Dawson, E., DeWitt, J., Eakins, A., & Wong, B. (2015). "Science capital": A conceptual, methodological, and empirical argument for extending bourdieusian notions of capital beyond the arts. *Journal of Research in Science Teaching, 52*(7), 922–948. https://doi.org/10.1002/tea.21227

Barman, C. R. (1997). Students' views of scientists and science: Results from a national study. *Science and Children, 35*(1), 18.

Betz, D. E., & Sekaquaptewa, D. (2012). My fair physicist? Feminine math and science role models demotivate young girls. *Social Psychological and Personality Science, 3*(6), 738–746. https://doi.org/10.1177/1948550612440735

Bodzin, A., & Gehringer, M. (2001). Breaking science stereotypes. *Science and Children, 38*(4), 36.

Buldu, M. (2006). Young children's perceptionsof scientists: A preliminary study. *Educational Research, 48*(1), 121–132. https://doi.org/10.1080/00131880500498602

Chambers, D. W. (1983). Stereotypic images of the scientist: The draw-a-scientist test. *Science Education, 67*(2), 255–265.

Christidou, V. (2010). Greek students' images of scientific researchers. *Journal of Science Communication, 9*(3), A01. https://doi.org/10.22323/2.09030201.

Christidou, V., Hatzinikita, V., & Samaras, G. (2012). The image of scientific researchers and their activity in Greek adolescents' drawings. *Public Understanding of Science, 21*(5), 626–647. https://doi.org/10.1177/0963662510383101

DeWitt, J., Archer, L., & Mau, A. (2016). Dimensions of science capital: Exploring its potential for understanding students' science participation. *International Journal of Science Education, 38*(16), 2431–2449. https://doi.org/10.1080/09500693.2016.1248520

Dickson, M., & McMinn, M. (2022). Children's perceptions of scientists and their work: The 'Draw a Scientist' test in the United Arab Emirates. *Public Understanding of Science, 31*(8), 1079–1094. https://doi.org/10.1177/09636625221096795

Emvalotis, A., & Koutsianou, A. (2018). Greek primary school students' images of scientists and their work: Has anything changed? *Research in Science & Technological Education, 36*(1), 69–85. https://doi.org/10.1080/02635143.2017.1366899

Farland-Smith, D. (2012). Personal and social interactions between young girls and scientists: Examining critical aspects for identity construction. *Journal of Science Teacher Education, 23*(1), 1–18. https://doi.org/10.1007/s10972-011-9259-7

Finson, K. D., Beaver, J. B., & Cramond, B. L. (1995). Development and field test of a checklist for the draw-a-scientist test. *School Science and Mathematics, 95*(4), 195–205. https://doi.org/10.1111/j.1949-8594.1995.tb15762.x

Francis, B., Archer, L., Moote, J., DeWitt, J., MacLeod, E., & Yeomans, L. (2017a). The construction of physics as a quintessentially masculine subject: Young people's perceptions of gender issues in access to physics. *Sex Roles, 76*(3–4), 156–174. https://doi.org/10.1007/s11199-016-0669-z

Francis, B., Archer, L., Moote, J., de Witt, J., MacLeod, E., & Yeomans, L. (2017b). Femininity, science, and the denigration of the girly girl. *British Journal of Sociology of Education, 38*(8), 1097–1110. 10.1080/01425692.2016.1253455.

Fung, Y. Y. (2002). A comparative study of primary and secondary school students' images of scientists. *Research in Science & Technological Education, 20*(2), 199–213. https://doi.org/10.1080/0263514022000030453

Godec, S., King, H., & Archer, L. (2017). The science capital teaching approach: engaging students with science, promoting social justice. Teaching Resource. Retrieved from https://discovery.ucl.ac.uk/id/eprint/10080166/1/the-science-capital-teaching-approach-pack-for-teachers.pdf.

Latu, I., & Mast, M. S. (2015). The effects of stereotypes of women's performance in male-dominated hierarchies: stereotype threat activation and reduction through role models. In *Gender and Social Hierarchies* (pp. 87–99). Routledge.

Hill, D., & Wheeler, A. (1991). Towards a clearer understanding of students' ideas about science and technology: An exploratory study. *Research in Science & Technological Education, 9*(2), 127–137. https://doi.org/10.1080/0263514910090202

Hillman, S. J., Bloodsworth, K. H., Tilburg, C. E., Zeeman, S. I. & List, H. E. (2014). K-12 students' perceptions of scientists: Finding a valid measurement and exploring whether exposure to scientists makes an impact. *International Journal of Science Education* (15), 2580–2595. https://doi.org/10.1080/09500693.2014.908264.

Leslie, S. J., Cimpian, A., Meyer, M., & Freeland, E. (2015). Expectations of brilliance underlie gender distributions across academic disciplines. *Science, 347*(6219), 262–265. https://doi.org/10.1126/science.1261375

Losh, S. C., Wilke, R., & Pop, M. (2008). Some methodological issues with "Draw a Scientist Tests" among young children. *International Journal of Science Education, 30*(6), 773–792. https://doi.org/10.1080/09500690701250452.

Mason, C. L., Kahle, J. B., & Gardner, A. L. (1991). Draw-a-scientist test: Future implications. *School Science and Mathematics, 91*(5), 193–198.

Mead, M., & Metraux, R. (1957). Image of the scientist among high-school students. *Science, 126*(3270), 384–390.

Milford, T. M., & Tippett, C. D. (2013). Preservice teachers' images of scientists: Do prior science experiences make a difference? *Journal of Science Teacher Education, 24*(4), 745–762. https://doi.org/10.1007/s10972-012-9304-1

OECD. (2014). *PISA 2012 Results: What students know and can do—student performance in mathematics, reading and science (Vol. I, Revised edition, February 2014).* OECD Publishing. https://doi.org/10.1787/9789264201118-en

OECD. (2019). *PISA 2018 Results (Volume I).* https://doi.org/10.1787/5f07c754-en.

Orlander, A. A. (2019). When "scary" science "just feels wrong": how the facts in a masculine fact-based debate couldn't stop science students' feminine feelings. *Cultural Studies of Science Education*, 1–21.https://doi.org/10.1007/s11422-018-9904-y.

Ozel, M. (2012). Children's images of scientists: Does grade level make a difference? *Educational Sciences: Theory and Practice, 12*(4), 3187–3198.

Reay, D. (2001). "Spice Girls", "Nice Girls", "Girlies", and "Tomboys": Gender discourses, girls' cultures and femininities in the primary classroom. *Gender and Education, 13*(2), 153–166.

Read, B. (2011). Britney, Beyoncé, and me–primary school girls' role models and constructions of the 'popular'girl. *Gender and Education, 23*(1), 1–13. https://doi.org/10.1080/09540251003674089

Reinisch, B., Krell, M., Hergert, S., Gogolin, S., & Krüger, D. (2017). Methodical challenges concerning the draw-a-scientist test: A critical view about the assessment and evaluation of learners' conceptions of scientists. *International Journal of Science Education, 39*(14), 1952–1975. https://doi.org/10.1080/09500693.2017.1362712

Rosenthal, D. B. (1993). Images of scientists: A comparison of biology and liberal studies majors. *School Science and Mathematics, 93*(4), 212–216.

Schinske, J. N., Perkins, H., Snyder, A., & Wyer, M. (2016). Scientist spotlight homework assignments shift students' stereotypes of scientists and enhance science identity in a diverse introductory science class. *CBE—Life Sciences Education, 15*(3).

Song, J., & Kim, K. S. (1999). How Korean students see scientists: The images of the scientist. *International Journal of Science Education, 21*(9), 957–977. https://doi.org/10.1080/095006999290255

Türkmen, H. (2008). Turkish primary students' perceptions about scientist and what factors affecting the image of the scientists. *Eurasia Journal of Mathematics, Science & Technology Education, 4*(1). https://doi.org/10.12973/ejmste/75306.

Dr. Martina Dickson holds a PhD in Physics from the University of London and an MA in Gender, Education and International Development from the IoE (UCL) in London. She has held a variety of teaching and advisory positions in Greece, Oman, Hong Kong and the UAE, and is currently a Professor in the Curriculum and Instruction Department at Emirates College for Advanced Education in the UAE. Her research interests include gender in education and science pedagogy, and she has led a number of research projects over the last decade, including studies on academic parenthood, female STEM leadership, gendered choices and perceptions of scientists and science careers.

Dr. Melissa McMinn has worked in in-service and pre-service teacher education for 15 years, and in post-graduate education for eight years. She is currently a Learning Designer for the University of Auckland, New Zealand. She holds a Master's of Education and a Doctorate of Philosophy in Mathematics and Science Education and has achieved the status of Senior Fellow (SFHEA) in recognition of her teaching and learning support in higher education. She is an active researcher and over the past decade has led projects in mathematics and science anxiety in university students and teachers, and learning environments research, and co-researched a number of projects including pedagogy in higher education, and children's perceptions of science and technology, among others.

Part II
Attitudes and Understanding in STEM

If one could imagine a continuum between affective and cognitive domains, beliefs would sit near the cognitive end. As discussed in Part I, beliefs are relatively stable and thus, harder to change. Anxiety, however, is primarily affective in nature and is often described as unpleasant emotions. While the terms anxiety and attitudes have been used both synonymously and separately, attitudes fall somewhere in the middle of the continuum, including both cognitive and affective components.

Both anxiety and attitudes towards learning affect motivation and the amount of effort one is willing to exert to learn. It is possible that this has a snowball effect; fatalistic attitudes and high anxiety in students can become a self-fulfilling prophecy. Students' attitudes and levels of anxiety and motivation towards STEM learning have great impact on the desire to select STEM courses beyond minimum requirements in secondary school and enrol in (and complete) STEM-related programmes in higher education. Such constructs will also influence the activities and interests that students may pursue beyond the classroom, such as taking an interest in their physical world, and are ultimately likely to affect student literacy in STEM subjects.

Given that attitudes and anxiety are likely to be more malleable than beliefs, and that both these outcomes, as well as motivation and achievement, have been related to aspects of the learning environment including teacher attributes and pedagogies, it is possible to improve these outcomes by making changes to the learning environment. It is therefore important to consider the learning environment and ensure teachers are well-equipped to guide students to positive outcomes.

Chapter 4
Changing Perceptions of the Learning Environment and Attitudes Towards Mathematics Through Inquiry-Based Learning: Girls in Middle School Classrooms in the UAE

Jennifer Robinson and Jill Aldridge

Abstract The study reported in this chapter was carried out with students in four all-girl middle schools in the Emirate of Abu Dhabi. The study took place during a large-scale education reform effort driven by the need to build a knowledge-based economy within the emirate. As part of the reform effort, inquiry-based learning (IBL) was mandated across all middle school mathematics classes to increase student engagement and their enjoyment of the subject. Given that attitudes towards mathematics are important determinants of whether students persevere and pursue further study and potential careers, the mandate held great promise. Despite the introduction of IBL across schools, after approximately five years teachers were implementing the approach with differing degrees of success, with some teachers struggling to implement the approach at all. This provided us, as researchers, with an opportunity to compare students' learning environment perceptions and attitudes towards mathematics in classes where IBL was being implemented effectively with those that remained largely traditional in terms of teaching. The study involved a mixed-method approach in which both quantitative and qualitative data were collected and analysed. Quantitative data were collected from 291 students using two surveys, one to assess perceptions of the learning environment and another to assess students' attitudes towards mathematics. Qualitative data, gathered from 27 students, were gathered to provide impressionistic tales aimed at contextualising the setting as well as giving causal explanations for the quantitative findings. Our results suggested that in classes with teachers exemplary in their use of IBL, students had statistically significant ($p < .01$), positive perceptions of their learning environment and attitudes towards mathematics. Qualitative information suggested that, in these classes, students were more engaged in their learning and had more agency. The nature of the IBL meant that students were able to use more real-life applications and, therefore,

J. Robinson (✉)
Emirates College for Advanced Education, Abu Dhabi, UAE
e-mail: Jennifer.Robinson@ecae.ac.ae

J. Aldridge
Curtin University, Perth, Australia

M. Dickson et al. (eds.), *Gender in STEM Education in the Arab Gulf Countries*,
https://doi.org/10.1007/978-981-19-9135-6_4

valued the tasks more. Further, IBL classes provided students with greater opportunity to carry out investigations. These findings are of significance to education providers as well as curriculum developers. Further, given the underrepresentation of girls in post-compulsory STEM education and related careers, compared to boys, our findings offer hope for education systems wishing to overcome this disparity. This chapter offers recommendations for stakeholders based on the benefits of these pedagogical approaches in UAE schools.

Keywords Inquiry-based learning · Learning environments · Mathematics education · Student attitudes

Introduction

The United Arab Emirates (UAE) has been undergoing comprehensive educational reform efforts for a number of years with a focus on moving from traditional didactic modes of education delivery to that of a student-centred pedagogy. An emphasis has been on promoting STEM (Science, Technology, Engineering and Mathematics) education and career pathways for students, particularly for girls. Traditionally girls have been underrepresented in STEM courses in higher education and career fields, where often high salaries are on offer (Anaya et al., 2021; OECD, 2020). The UAE is seeking to move from a reliance on oil and gas to a knowledge-based economy and the need for citizens to be prepared for this shift has been identified as an essential element of the change. The importance of promoting STEM careers as options for girls has been recognised, and strategic thinking concerning how to implement this is underway.

The research discussed in this chapter is part of a larger study (Robinson, 2020) that explores the findings of utilising inquiry-based learning (IBL) in public schools in Abu Dhabi and girls' perceptions of their learning environment and attitudes towards mathematics (Robinson & Aldridge, 2022). Students' perceptions of the learning environment were used as a process criterion to assess the effectiveness of the pedagogical changes utilised as part of the educational reform in middle school mathematics classes. The importance of investigating girls' attitudes towards mathematics was paramount, considering the intention to encourage girls to pursue further study and careers in STEM.

The use of a comparison in this research where IBL was implemented effectively and where it was not, allowed for a contrast to define the differences in many aspects of the classroom and to suggest solutions and a way forward for individual teachers and for the reform effort as a whole. IBL was implemented through the use of explorations, open-ended inquiry tasks with students working in collaborative groups. Students were required to solve authentic or 'real' problems (Akaygun & Adadan, 2021) in a cyclic manner with students proceeding, in general, through the following steps: questioning, planning, researching, analysing, concluding and reflecting (Kogan & Laursen, 2014).

The research sought to investigate the following two objectives.

1. To investigate whether learning environment perceptions and attitudes differed for students taught by teachers exemplary in the use of explorations and those who were not.
2. To investigate reasons for differences in the perceived learning environment and attitudes towards mathematics for students taught by teachers exemplary in the use of explorations and those who were not.

Background to the Study

Inquiry-Based Learning

A significant focus of the educational reform efforts in Abu Dhabi included the shift from traditional teaching methods to IBL techniques, requiring teachers to use exploration projects based on real-world contexts. The use of IBL is widely viewed as an authentic approach to teaching and learning mathematics (Amaral et al., 2002). For the context of this study, IBL was defined as a "student-centered pedagogy that uses purposeful, extended investigations, set in the context of real-life problems, as both a means for increasing student capacities and as a feedback loop for increasing teachers' insights into student thought processes" (Supovitz et al., 2000, p. 332).

The process of an inquiry has general steps which are not normally performed in a linear fashion but, rather, follow a cyclic process that involves: questioning, planning, researching, analysis, concluding and reflecting (Chapman & Heater, 2010; Jansen, 2011; Singer & Moscovici, 2008). In the context of mathematics, students are required to solve authentic or 'real' problems (Makar, 2007) by beginning with a big question or idea that they wish to investigate and ending with an answer that generates more questions to explore. Collaboratively, students plan their process and spend time in research through multiple sources, which may include print, electronic and people sources (Chapman & Heater, 2010). The investigative stage may also include the use of mathematical experiments, simulations, or the use of a survey to gather data for answering the big question. At this point in the cycle, the teacher may need to explain content and information that students may not have had the opportunity to gather or understand for themselves. An important aspect of an investigation is the development of the ability for students to filter information and to analyse what has been discovered. Next, students are required to use mathematics to explain their discoveries and to apply this new knowledge in unfamiliar contexts. This step supports students in forming conclusions to answer their big question. The final step of the cycle is when students present their findings, reflect on their conclusions and seek to consider where the inquiry goes next. This then informs the engagement in the next big questions and the cycle continues (Wu & Hsieh, 2006). Makar (2007, p. 48) states that "the goal of inquiry is both knowledge-building and building understanding of the processes of knowledge-building, that is, learning how to learn".

As a teaching approach, IBL can encourage students to think critically (Marshall et al., 2010); reason (Staples, 2007); develop deep understandings (Makar, 2007); and foster positive identities regarding mathematics (Staples, 2007) while experiencing opportunities to link learning to their real lives (Gholam, 2019). This approach is far removed from the traditional methods used in the UAE that have previously relied on the memorisation of facts (Marshall & Horton, 2011). Of particular note, the inquiry approach encourages students to reflect critically (Leikin & Rota, 2006) and be intellectually engaged in the task (Oliver, 2007).

For these reasons, the use of IBL was promoted as part of the pedagogical shift in the educational reform for mathematics. The intention was to utilise this approach as a means of developing students critical thinking, creativity, communication and collaboration skills, required by the twenty-first century learners (Abu Dhabi Education Council, 2013).

Learning Environments

Students spend a substantial amount of time within classrooms and most of the focus is on their academic achievement scores. While this is an important aspect of school life, the environment in which students are learning is an essential component and cannot be ignored. The learning environment has two elements: the physical component, including the room, furniture and resourcing, and the psychosocial aspect, incorporating teacher–student interpersonal relationships and interactions between students (Fraser, 2015). This research focused on the psychosocial learning environment, including the culture, atmosphere or ambience where learning occurs (Fraser, 2012).

Consideration of the learning environment is important to learning in general, as a positive learning environment has been shown to promote positive outcomes for students, such as an increase in students' motivation and a decrease in apprehension for students (Ellis, 2004). Students' perceptions of the learning environment are related to academic achievement (Anderson et al., 2004; Chionh & Fraser, 2009; Wolf & Fraser, 2008) and, when considering the factors impacting achievement, it was found that a focus on competition and a comparison of student ability resulted in a decrease in achievement (Wang & Holcombe, 2010). Conversely, positive perceptions of the learning environment have been shown to increase enjoyment and promote positive student attitudes, which is a specific focus of the current study (Aldridge, Fraser, et al., 2012a, 2012b). In the context of an IBL setting, the learning environment is unique and its impact on student learning is an important consideration. The nature of IBL in the classroom is that it requires an environment that allows for student discussion and discourse, opportunities for students to generate several solutions and for students to make decisions and to provide justification for these decisions (Bruder & Prescott, 2013). For this reason, assessing the learning environment to address the impact of IBL in the classroom was a meaningful measure.

This study focused on students' perceptions of six aspects of the learning environment: personal relevance, critical voice, student negotiation, shared control, investigation and involvement. Personal relevance assessed the extent to which opportunities for students to have a personal meaningful connection to their learning are provided (Priniski et al., 2018). Past research has found that when lessons are personally relevant to students, there is increased engagement and enjoyment (Aldridge, Afari, et al., 2012a, 2012b; Priniski et al., 2018). Traditional mathematics classrooms in Abu Dhabi were places where mathematics was taught from a theoretical perspective with little application to real-world experiences and almost no content relevant to the life of a student growing up in a rural or urban context in a desert area in the Middle East (Gaad et al., 2006) Assessing the personal relevance of the educational innovation of exploration approaches to teaching and learning, therefore, was important to determine whether students found them to be relevant to their lives outside of school.

Critical voice, from a critical theory perspective, is about giving students opportunities to question teacher's pedagogical methods and activities and empowering students to be able to discuss any restrictions that they encounter in their learning. Giving students a voice empowers them and makes them feel that they belong, are valued and that their contributions matter (Bain, 2010). Traditionally, mathematics teaching in the UAE has involved the teacher lecturing directly from a government-provided textbook and students completing all of the exercises contained within (Gaad et al., 2006). Assessing the extent to which teachers allowed students to have a critical voice in the classroom was considered to be a good indicator of change in pedagogy towards lessons that are more student centred.

The use of student negotiation promotes opportunities "for students to explain and justify to other students their newly developing ideas, to listen attentively and reflect on the viability of other students' ideas and, subsequently, to reflect self-critically on the viability of their own ideas" (Taylor et al., 1997, p. 4). In this context, explorations are completed in groups of between two to four students, with students being required to present ideas to the other group members and come to a consensus about how to proceed. Therefore, assessing the extent to which students' perceived opportunities to negotiate with each other was an important feature of the learning environment.

Shared control in classroom settings is about the "extent to which students are invited to share control of the learning environment with the teacher, including the articulation of their own learning goals, design and management of their learning activities and determining and applying assessment criteria" (Ozkal et al., 2009, p. 72). There is a need for learners to construct their own knowledge through the use of shared control between teachers and students (Sultan et al., 2011) as it makes learning more enjoyable with positive effects on learner motivation (Partin & Haney, 2012). Shared control was selected to assess whether students felt that they were able to help the teacher to decide what activities were best for them in the context of an exploration inquiry task.

The use of investigation in the classroom is the "extent to which skills and processes of inquiry and their use in problem-solving and investigation are emphasised" (Dorman, 2008, p. 183). Mathematical investigation is based on the pedagogical belief that students learn best when they have opportunities to be active learners and to construct personal understandings of mathematical concepts (Alt, 2018). Investigation was chosen for use in this study to explore the extent to which students believe that they have opportunities to investigate while completing an exploration inquiry task. Given that investigation is essential for effective explorations this construct was important.

Involvement is described as “the extent to which students have attentive interest, participate in discussions, do additional work and enjoy the class” (Dorman, 2008, p. 183). Students who are involved in their lessons exhibit different behaviours to their peers through use of body language, verbal participation and social interactions. They are more willing to attempt mathematical problems, ask questions when clarification is required and are more active in their learning (Nebesniak & Heaton, 2010). Including involvement in the study provided the opportunity to assess the extent that students feel involved or able to participate within the classroom. In the context of this study, involvement included discussions about the work or involvement in collaborative activities in which students work together towards a common goal.

The use of learning environments as a process criterion to assess the effectiveness of an educational innovation, such as the introduction of IBL, has been well documented (Fraser, 2018). “Research findings have consistently shown that students’ and teachers’ perceptions of important social and psychological aspects of the learning environments really matter in terms of educational outcomes” (Koh & Fraser, 2014, pp. 158–159), emphasising the need to consider class room learning environments when implementing educational reforms. A wealth of research has focused on assessing the effectiveness of innovative pedagogies, such as the introduction of IBL as evidenced in this study, by examining the impact of the new methodologies on students’ perceptions of the learning environment (see, for example, studies by Afari et al., 2012; Baeten et al., 2013). Using the previous research as a basis, this study drew on and extended the research to the unique setting of Abu Dhabi classrooms to examine the impact of the introduction and implementation of IBL on mathematics learning.

Attitudes

When considering mathematics teaching and learning in schools and the subsequent promotion of student participation in higher education and mathematics-related careers, the importance of attitudes is an integral element. The concept of attitude was originally discussed by Herbert Spencer in 1862, and since then many researchers have defined attitudes from differing contexts and points of view. There is consensus, however, that attitudes are made up of three components: the cognitive component that describes the knowledge, beliefs and ideas about an object; the affective component which describes the feeling about an object in terms of like or dislike and the behavioural component which describes a tendency-towards-action (Breckler, 1984). In the context of mathematics education, an attitude to mathematics has been defined as "an aggregated measure of a liking or disliking of mathematics, a tendency to engage or avoid mathematical activities, a belief that one is good or bad at mathematics, and a belief that mathematics is useful or useless" (Neale, 1969, p. 632).

Within this study, three aspects of attitude research were focused on: one related to the cognitive component (self-efficacy) and two related to the affective component (enjoyment of mathematics classes and task value). Enjoyment in mathematics has been shown to foster problem-solving, promote resilience and support a student's individual self-regulation (Pekrun et al., 2002). Enjoyment is considered to be a factor in impacting students' behaviour and engagement in the classroom (Hidi & Renninger, 2006) and extensive research has shown a positive relationship between enjoyment of mathematics and achievement (see, for example, Ahmed et al., 2013; Goetz et al., 2012). Self-efficacy influences the decisions students make concerning their choice of activities and tasks and can determine whether they are willing to engage with activities that they may consider beyond their capabilities (Bandura, 1977). Students with strong self-efficacy will be more willing to persevere with a task they find difficult and expend additional energy and determination to succeed (Choi, 2005). Task value refers to the perception by a student of the usefulness of a particular activity and impacts their willingness to persevere if they deem the outcome to be important. Task value is a strong predictor of achievement-related outcomes (Acee et al., 2018) and can impact choices concerning continuing education and potential course and career selections (Eccles et al., 1983).

Considering the impact of positive attitudes in establishing pathways for girls into mathematical programmes in higher education and STEM career fields, a focus on attitudes towards mathematics was pertinent (Riegle-Crumb et al., 2019). Bearing in mind that task values, an attitudinal component in this study, have been shown to strongly relate to career choices and aspirations (Lauermann et al., 2017; Lazarides & Watt, 2015) and so are a strong factor for its inclusion. When comparing to boys, girls tend to have lower levels of enjoyment of mathematics (Gaspard et al., 2015), thus reporting lower levels of interest in pursuing mathematics-intensive careers (Lazarides & Lauermann, 2019) further explaining that enjoyment of mathematics is a key factor in determining whether girls will pursue a STEM career (González-Pérez et al., 2020). Research strongly supports the connection between positive learning environments and student attitudes. Attitudes and enjoyment towards mathematics do not remain in fixed parameters. Instead, attitudes can be formed and changed, both directly and indirectly, through experiences within the learning environment (Simonson & Maushak, 2001; Vandecandelaere et al., 2012). By altering the teaching methodology and climate within the classroom, through such things as giving feedback, coaching and supporting students in self-regulation, students' enjoyment of mathematics can be improved (Vandecandelaere et al., 2012). Perceptions of the learning environment have been shown to be related to students' attitudes and enjoyment of the subject (Aldridge, Fraser, et al., 2012a, 2012b; Wolf & Fraser, 2008).

Through the establishment of IBL processes in classes, the three attitudinal aspects, enjoyment of mathematics, self-efficacy and task value, have been shown to increase. The use of IBL as a pedagogical approach in the classroom has been found to increase students' enjoyment (Camenzuli & Buhagiar, 2014) through making the content more interesting (Bruder & Prescott, 2013). In addition, past research has indicated that the use of IBL in classes promotes students' self-efficacy (Laine et al., 2017) and where students have opportunities to participate in inquiry classes, they display more belief in their capabilities to succeed, therefore encouraging them to consider further study and career options (Kang & Keinonen, 2017). As well as the benefits shown in self-efficacy, Heindl and Nader (2018) found the use of IBL techniques increased the value students placed on the tasks they completed. In a study set in a primary mathematics class, the contextual nature of inquiry approaches allowed students to solve authentic problems, increased their interest and encouraged a greater sense of value in the tasks assigned (Fielding-Wells et al., 2017). The research discussed in this chapter outlines how the use of IBL in Abu Dhabi mathematics classes promoted positive attitudes towards mathematics and indicated positive student perceptions of the learning environment. This extends the findings of previous studies and provides a unique view of Abu Dhabi education, showing a unique and important gap in the current research.

Methods

In this section, the processes and methods used in the collection and analysis of the data are described. These include the theoretical framework, sample, instruments and analysis completed.

Theoretical Framework

The study reported in this chapter involved a mixed methods design. The utilisation of mixed methods allowed for a more useful investigation into the research topic rather than limiting to either research type on its own. The administration of mixed methods was advantageous as it provided a means to offset any weaknesses found within the individual processes involved in quantitative and qualitative research. Difficulties within quantitative research are often due to the inability for the context or the voices of the participants to be heard, whereas in qualitative research, it can be difficult to generalise findings due to small sample sizes. In addition it can be considered as deficient due to the researcher's personal interpretations (Creswell & Plano Clark, 2017). In order to reduce these issues as much as possible, the utilisation of mixed methods research was chosen, to allow for both statistical and thematic analysis and personal interpretations to provide insights into the research objectives.

Pragmatism was drawn on as an overarching paradigm for the mixed methods approach used in this research. Classical theorists, Peirce (1878), James (1907, 1995) and Dewey (1920, 1948), first discussed the pragmatic paradigm and, by general consensus, it focuses on practical applications, rather than on what can be considered unquestionably true or tangible (Weaver, 2018). In this study, when differing approaches appeared to be philosophically inconsistent, the pragmatic paradigm guided the research design.

The two phases of the study utilised differing paradigms in each. Surveys were used in the first phase in the collection of quantitative data. The second phase involved gathering qualitative data through lesson observations and focus group interviews with students. Each phase involved the use of a different paradigm, as discussed below.

The first phase of the research process drew on the paradigm of post-positivism that guided the methodology using a quantitative design and the use of surveys as data collection instruments. Guided by the post-positivist design, an important aspect of the data analysis was to show that the findings were valid and not based on the judgement of the researcher, through the use of established procedures (Firestone, 1987). Included in this, were the details of the components within the sample, the methods utilised to collect the quantitative data, the statistical processes used to analyse the data and the findings of the research. This allowed for limited objectivity from the researcher with the tone of the analysis as scientific and precise.

The second phase of the study, in which qualitative data was collected, involved an interpretivist approach. The purpose of interpretative research is to understand people's experiences, and so the data collection occurred within the natural setting of the students, the classroom. The research objectives for the qualitative phase were open-ended and aimed to support the findings of the quantitative phase of the study. Due to the subjective nature of the qualitative data collection, through the use of lesson observations and focus group interviews, it was essential that the researcher collected the data. Analysis of the data required the use of rich descriptions of the procedures and resulting findings including quotes and use of the quantitative data to support the findings. The use of the interpretative stance meant that the role of researcher was subjective and so the rhetoric utilised in the description of the findings was in the first person and personalised, including information on any biases and experiences in the research (Ponterotto, 2005).

Sample

Quantitative Data

The sample was drawn from four girls' middle schools (grades 6–9) in the environs of Abu Dhabi city in the United Arab Emirates. All schools were public schools where the medium of instruction for the majority of the classes was Arabic, thus making the students English as an Additional Language (EAL) learners. Middle schools were chosen due to their extended experience with IBL as part of the educational reform efforts and each individual school was chosen using a purposive sampling strategy based on the selection of individual teachers within. Two teachers were identified as exemplary practitioners of IBL and two were not. Teachers were selected for the study using a purposive sampling strategy. Criteria for teachers exemplary in the use of explorations were developed by the researcher, based on the requirements of the inquiry approach as outlined in the exploration rubric and assessment documentation, and these were used to select teachers for this study. Teachers were not informed whether they had been selected as a teacher exemplary in the use of inquiry or not. The criteria were related to six areas and are summarised below.

- Demonstration of skills required in the explorations rubric and framework through teaching and learning experiences as well as assessment opportunities.
- Opportunities for students to explore new concepts.
- Skilful questioning including open-ended questions and those to stimulate thinking.
- Opportunities for students to reflect on processes and products and to discuss limitations and further recommendations.
- Time for collaboration within lessons.
- Use of an extended community of learners and experts outside of the school.

Selection of the teachers was made through a recommendation process from subject-specific expert mathematics education advisors using the criteria. Mathematics education advisors were employed by the education department as part of the reform effort to mentor and coach teachers on matters of curriculum, assessment and pedagogical practice. The education advisors were well placed to make sound recommendations of teachers as they were based in the schools, observed lessons regularly and knew the teachers and their levels of expertise very well. Data collection occurred well into the school year to allow students' perceptions of their mathematics classes to reflect that of the class they were in and not to be impacted by perceptions of the learning environment and attitudes towards mathematics from previous academic years. All teachers were currently employed in schools and students remained in their existing classes. Surveys were administered to 291 students in 12 mathematics classes (77 grade 7 students, 141 grade 8 students and 73 grade 9 students).

Qualitative Data

Lesson observations occurred in six classes, three with a teacher exemplary in the implementation of explorations and three with a teacher who was not. Both teachers had classes that had previously participated in the quantitative phase of the data collection. After each observation, focus group interviews were conducted using a purposive sampling method. Two criteria were used to establish the interview groups. First, in order to keep the number of adults to a minimum within the groups to allow students to speak freely, students were chosen based on their ability to communicate in English to ensure a translator was not required. The second criterion was to ensure students with a range of attitudes and abilities were included in the sample. Using the teachers to assist, students were selected who had a range of academic results in mathematics and a variety of effort grades.

Each of the six focus groups contained up to four or five students, with three focus groups involving students taught by teachers exemplary in the use of explorations and three made up of students taught by teachers non-exemplary in the use of explorations. There were 27 students interviewed in total.

Instruments

Learning Environment in Inquiry Survey (LEIS)

Two surveys were developed from existing validated instruments in order to meet the needs of the unique context of this specific study in the educational reform in Abu Dhabi. The Learning Environment in Inquiry Survey (LEIS; Robinson & Aldridge, 2022) utilised four scales (personal relevance, critical voice, student negotiation and shared control) from the Constructivist Learning Environment Survey (CLES; Taylor et al., 1997) and two scales (involvement and investigation) from the What Is Happening In this Class? Questionnaire (WIHIC; Aldridge et al., 1999). Students responded to the items using a five-point frequency-response scale of almost always (=5), often, sometimes, seldom and almost never (=1). Using the LEIS provided students with the opportunity to express their opinions about the learning environment in their mathematics classes.

Student Attitudes Towards Mathematics Survey (SATMS)

A second instrument, the Student Attitudes Towards Mathematics Survey (SATMS; Robinson & Aldridge, 2022), was developed to assess students' attitudes towards learning in mathematics, particularly in the inquiry-based learning context. The survey incorporated one scale (enjoyment of mathematics classes) from the Test of Mathematics Related Attitudes (TOMRA; Spinner & Fraser, 2005) and two scales (self-efficacy and task value) modified from the Student Adaptive Learning Engagement in Science questionnaire (SALES; Velayutham et al., 2011). Students responded to each of the items using a five-point frequency-response format of almost always (=5), often, sometimes, seldom and almost never (=1).

Lesson Observations

The purpose of the lesson observations was to collect data of a qualitative nature to add richness to the quantitative data and, where pertinent, to provide causal explanations for the survey results. All classes observed were intended to be lessons where inquiry techniques were used as the main pedagogical approach. The content being taught was at the discretion of the specific teacher according to the scope and sequence directed by the curriculum from the education department. All teachers participating in the study believed they were implementing the requirements of the educational reform in mathematics through the use of explorations utilising an IBL approach. All observations were predominantly non-participant and, therefore, the researcher sat at the back of the room however, during activities the researcher was able to move around the classes to discuss the lesson with students (where appropriate). Any conversations between the researcher and students involved open-ended questions to elicit responses from students allowing them to describe what they were doing and how they felt about the work that they were completing.

During the lessons, the researcher recorded observations using field notes by focusing on topics that linked with the survey instruments. Each topic had guiding questions to help focus on aspects of the learning environment and attitudes that would provide explanations and insights into the quantitative results. The field notes were later typed for analysis and divided into the topics mentioned above. By structuring the field notes in this way, a direct link was provided to the topics explored in the quantitative data collection. While the researcher did not force or manufacture comments for each of the topics, they were intentional in observing the class through the lens of each area of focus and sought to write notes according.

Focus Group Interviews

After each lesson observation, the researcher facilitated focus group interviews with the aim to further understand aspects of the lesson and general experiences in their mathematics classes. The focus group interviews sought to add an important dimension to the lesson observations as both aimed to provide causal explanations and deeper insight into the quantitative survey data. Focus groups were chosen as, instead of relying on a formal mode of delivery where the interviewer asks questions of the group and participants reply in turn, the emphasis is on the discussion and interactions between group members (Morgan, 1988).

The focus group interviews were semi-structured, guided by the same topics that were used during observations and recorded using field notes. The use of semi-structured interviews ensured that all-important topics were covered but allowed for discussions to follow lines of interest that were observed during the lesson. An interview schedule based on the key topics was used to guide questions and probes for further discussion and to ensure consistency across the various focus groups.

Analysis

A one-way multivariate analysis of variance (MANOVA) was used to investigate whether there were differences in students' perceptions of the learning environment and attitudes towards mathematics with teachers who were exemplary and not exemplary in the use of explorations. For MANOVA, the individual student was used as the unit of analysis with the six learning environment scales and the three attitude scales representing the dependent variables and the type of teacher (exemplary/non-exemplary) representing the independent variable. Wilks' lambda, a test statistic in MANOVA, was calculated to assess whether the group differences were statistically significant. Wilks' lambda is a measure of the percent variance in dependent variables that are not explained by differences in levels of the independent variable (Field, 2016). To interpret the results of the ANOVA, the F statistic was used to measure how much the model improved the prediction of the outcome compared to the level of inaccuracy in the model.

To examine the extent of the differences between the student perceptions of the learning environment and attitudes towards mathematics for teachers exemplary in the use of explorations and those who were not, Cohen's effect sizes were calculated. These effect sizes were calculated in terms of the differences in means divided by the pooled standard deviation as suggested by Thompson (2001).

Impressionistic Tales

The purpose of analysing the lesson observations was twofold: to tell a story that set a context for the reader and to provide ease with comparing the two scenarios, classes where IBL was implemented in an exemplary manner and classes where it was not. The lesson observations were described using impressionistic tales, which are a way of providing an insider's view of the classroom allowing readers to feel part of the experience and use the five senses to fully immerse themselves in the story and environment (Van Maanen, 1982).

Two tales were written to represent the two classroom contexts, where each tale was representative of three lesson observations describing events and quotations. Rather than providing three similar tales, pertinent aspects from the three lessons observed of teachers exemplary in the use of IBL were extracted and described in one impressionistic tale. This was also the process undertaken for the three lessons observed of teachers non-exemplary in the use of IBL, to generate one impressionistic tale.

Thematic Analysis

Thematic analysis is "a method for identifying and analysing patterns in qualitative data" (Clarke & Braun, 2013, p. 120). Merton (1975) first named thematic analysis as an approach for the analysis of qualitative data; however, many different versions have been proposed since. The focus of this research was on the six phases of thematic analysis as described by Clarke and Braun (2013). The first phase, familiarisation with the data, involved the researcher becoming cognizant with the data through reading and re-reading of the field notes. These were then converted into proper cohesive sentences after the collection process, from the lesson observations and focus group interviews. As recommended by Creswell (2008), the field notes were read using preliminary exploratory analysis by studying the data to gain a general sense of it, noting ideas and planning the organisation and making decisions concerning whether additional data was required. A process of progressive focussing (Parlett & Hamilton, 1976) was used to start with a wide-angle lens, then narrowing the review and allowing key ideas to emerge. The second stage was coding, by dividing the data in text form into pieces, labelling with codes, assessing for overlap or redundancy and generating themes incorporating a group of related codes. The third stage involved searching for themes by grouping related codes into a theme, by establishing coherent and meaningful patterns relevant to the research question to determine each theme. Fourth, the review of themes involved reflecting on whether the themes told a convincing and compelling story about the data and were suitable to describe the coded extracts in addition to the full set. At this point, some themes were collapsed into one. The fifth phase involved defining and naming the themes. This step ensured the name encompassed and summarised the key idea and told an appropriate story for the theme. Consideration was placed on how the theme fits into the overall picture of the data. The final phase was the writing-up which "involves weaving together the analytic narrative and (vivid) data extracts to tell the reader a coherent and persuasive story about the data and contextualising it in relation to existing literature" (Clarke & Braun, 2013, p. 121), where for this research it involved impressionistic tales, commentaries and the individual themes.

Results

In the first research objective, the intention was to investigate the differences between students taught by teachers who were exemplary in their use of explorations and those who were not in terms of their perceptions of the learning environment and attitudes towards mathematics in the context of an inquiry-based exploration approach in their classrooms. The findings of the analysis of these differences are described in the following section.

Differences Between Classes

The differences between students' perceptions of the learning environment for teachers exemplary in the use of explorations and teachers who were not are reported in Table 4.1. The results include the average item mean, average item standard deviation, effect size and MANOVA results for each learning environment (LEIS) and attitude (SATMS) scale.

An examination of the average item means, reported in the left-hand columns of Table 4.1, indicates that the average item means were higher for students in classes that were taught by teachers exemplary in the use of explorations than their counterparts who were not.

Table 4.1 Average item mean, average item standard deviation and difference (effect size and MANOVA with repeated measures) between students with teachers exemplary and non-exemplary in the use of explorations for the LEIS and SATMS

Scale	Average item mean		Average item standard deviation		Difference	
	Exemplary	Non-exemplary	Exemplary	Non-exemplary	Effect size	*F*
LEIS						
Personal relevance	3.75	3.43	0.79	0.76	0.41	12.14**
Critical voice	3.80	3.58	0.94	0.90	0.24	3.90**
Shared control	3.60	3.28	1.03	0.97	0.32	7.18**
Student negotiation	3.92	3.71	0.78	0.85	0.26	8.28**
Involvement	3.99	3.71	0.78	0.85	0.34	8.38**
Investigation	4.01	3.69	0.74	0.83	0.41	11.79**
SATMS						
Enjoyment	3.66	3.03	1.27	1.13	0.52	20.26**
Self-efficacy	4.21	3.70	1.27	1.13	0.42	25.17**
Task value	4.01	3.57	0.97	1.04	0.44	13.80**

N = 139 students in classes with teachers exemplary in the use of explorations and 152 students in classes with teachers who were not

$^{**}p < 0.01$

Effect size Cohen's d is defined as the difference between the two means (exemplary and non-exemplary) divided by the standard deviation

The results reported in Table 4.1, indicate that these differences were statistically significant ($p < 0.01$) for all six LEIS scales and three SATMS scales at the 99% significance level. For the LEIS, F values ranged from 3.90 (for critical voice) to 12.14 (for personal relevance). F values for the SATMS ranged from 13.80 (for task value) to 25.17 (for self-efficacy). That is, for all six LEIS scales and three SATMS scales, students in classes taught by teachers exemplary in the use of explorations had statistically significantly more positive views of the learning environment and attitudes towards mathematics when compared to the perceptions of students who were not.

The effect sizes, also reported in Table 4.1, ranged from 0.24 to 0.41 for the different LEIS scales, which are considered to be small to medium in effect (Cohen, 1988; Sawilowsky, 2009). The effect size for the differences between the two groups ranged from 0.42 standard deviations, for the self-efficacy scale, to over half a standard deviation (0.52 standard deviations) for the enjoyment of mathematics scale. These results were considered to be medium to large in effect (Cohen, 1988; Sawilowsky, 2009).

Causal Explanations for the Quantitative Findings

To meet the second research objective, impressionistic tales were developed to describe the lesson observations completed during the qualitative phase of data collection. One of these reflects the classes of the teacher exemplary in the use of explorations, and one reflects the classes with the teacher who was not.

Impressionistic Tale 1: Observation Field Notes from Classes with Teachers Exemplary in the Use of Explorations

As the teacher and I walk to her classroom, she explains that the lesson I am observing involves an exploration task titled "How long would it take to get to Pluto?". The task is part of a unit of work concerning the use of scientific notation and students have been working on their exploration for a number of lessons already. The room is not particularly large with the desks arranged in groups of between four and six. Even though it is winter, the air conditioning is pumping out cool air however, the room does not feel cold due to the student work lining the walls and the colourful resources on desks and stored throughout the room. Soon after we arrive, the students commence streaming into the room, shaking the teacher's and my hand as is usual practice. The students are very welcoming, helping me to arrange a desk and chair towards the back of the room. There are 27 grade 9 female students who appear to understand the norms of the classroom by quickly settling into seats and preparing for the class.

The teacher commences the lesson by welcoming the students and they reply in unison in Arabic. It is then time to introduce me to the girls and this is completed in English. I feel happy to be there with all students smiling and welcoming me. At this point, the teacher draws the students' attention back to herself and they quickly resettle. The introduction is given in a mixture of Arabic and English—partially switching between languages and partially translating. This is for the benefit of both me and the students. The reform requires teachers to transition more and more to the use of English in mathematics classes however, not all students are at the level where they can cope with this. Students are asked to choose someone from their group to provide an update on their progress with the exploration. A student from one group reports that they have calculated that it will take 19 years to travel to Pluto. The teacher asks how they feel about that length of time by getting them to calculate the age they would be when they returned. The consensus from the group is summarised by their spokesperson who states she wouldn't want to travel as she would miss out on years when she was young and so many experiences that couldn't be replaced. After all the groups have had an opportunity to explain where they were up to with the task, the teacher asks them to continue their work. It is interesting to watch the girls negotiate with each other and, through collaboration, to commence work—some by going through their calculations from previous days, others by collecting a presentation paper from the teacher's desk. All students appear to be quickly engaged in the task.

At this stage of the lesson, it is appropriate for me to move around the room and hold discussions with the various groups. It is interesting to note that all groups are eager to have me join them and to answer my questions. I am fortunate that any language barriers are quickly overcome by at least one student within each group having a strong grasp of English. A student in one group explains that they are in the middle of the exploration task having already been given the big question, deciding on their chosen mode of transport, and have used several lessons to research and perform calculations as part of their analysis to try to answer the question. They have also collaborated in deciding how they intend to share and present their findings. I ask a group if they enjoy learning this way and they respond, "We like this class, it's fun", "We like to work together" and "We like to learn about the planets and science."

Joining another group, we quickly begin discussing their approach to the task and they explain that they are using a commercial plane to travel to Pluto as it has many benefits including space to move around without restrictions. One girl in the group explains that they have been completing additional research on the topic and she states "We researched about NASA on the computer. They went to Pluto in 2006 and got there in 2015. I hope to be an astronaut." With the focus in the UAE on increasing interest in STEM fields for students, it is an exciting comment to hear.

As I move around the room and throughout the lesson, I note the presence of clear classroom norms of behaviour and routines. Students are orderly in arriving to the class, during transitions between whole class and group activities, and when collaborating and discussing with each other and the teacher. Noise levels within the room stay within a comfortable range where students are easily able to hold conversations with each other, but also without disturbing those around them. It is interesting to observe the teacher as well as she moves around the room. Initially, she starts the lesson at the front of the room, but once students have commenced group work, she moves to each group in turn, discussing, questioning, prompting, and encouraging when necessary. After asking one group where they are up to, she asks "do you think your final answer is sensible?". After allowing for some thinking time, a student responds, "we weren't sure if we were correct, so we talked to another group and they chose a slower plane and got a bigger answer for the time taken so we think our result sounds right". This leads to a discussion about identifying limitations in the inquiry design and processes followed and any issues with final results.

When visiting one group, I ask how the girls feel about working together and learning in this way. "I enjoy learning about math and science together. We get to research and discover new facts and then explain it to our friends in the class", one student immediately explains to me. Another student interjects, "We also get to choose how we will learn and how we will present to the class. We feel like we can choose the way."

As the time for the lesson is drawing to a close, the teacher gets the attention of the girls, and they all stop and listen. She then reminds them of the next tasks and asks them to commence tidying up. They immediately put scissors, marker pens and glue back into the baskets on each set of desks, while some girls roll up their presentation posters and put them by the teacher's desk for safe keeping till the next class. Once order is restored, the students file from the room with some thanking me for visiting as they leave. One student, with a large grin on her face, asks me to return soon.

In this impressionistic tale, a classroom is described where the busyness of students, the work they produced and the conversations occurring between students and between the teacher and students suggested the students were engaged with and enjoying the lesson. The tale suggests that the use of inquiry-based learning as both a teaching and assessment tool provided opportunities for students to relate the mathematical content to their own lives.

The lack of behavioural issues suggested the students were busy with the tasks and engaged in the work they were completing. The students were enjoying the lesson and were willing to talk about how much they liked learning in this way. Throughout the lesson, the students actively discussed the given task and worked together to develop the product.

In the next impressionistic tale, a lesson with a teacher whose use of explorations was not exemplary is described. In this tale, the teacher sought to teach using inquiry-based learning techniques; however, she was not successful in effectively planning and implementing an inquiry approach to teaching, learning and assessing. She had received the same level of professional development and support as the teachers exemplary in the use of inquiry; however, her lessons were different as shown in the next narrative.

Impressionistic Tale 2: Observation Field Notes from Classes with Teachers not Exemplary in the Use of Explorations

Early on a foggy morning, I head out of Abu Dhabi City to visit a school set in a small desert community. Once navigating through the thick fog, I easily find parking at the school, sign in with security, collect my visitor badge, and make my way to the principal's office following directions from reception staff. The principal is welcoming and invites me to visit anytime. Soon after I meet with the teacher whose lessons I have organised to observe, and she explains the plan for the class. The students have been working on a trigonometry unit of work and today they are using inquiry techniques to develop understanding of solving real-life problems involving right-angled trigonometry. The teacher explains that the lesson is directly after the break time and so although it is scheduled for 10.50am, she recommends I arrive at 10.55am as the students are expected to take a little extra time returning from their break.

When I arrive, the classroom is still empty, so I quietly wait outside for everyone. By 11am the teacher and most of the students are in the class and settling at desks. I quickly arrange a desk and chair for myself at the back of the classroom and use the time before the lesson commences to review the room itself. The air conditioning units are blasting freezing air around the room, but I seem to be the only person who is cold. The walls are painted white and there are no additional decorations—posters or student work so this adds to the sense of coolness. There is limited furniture in the room, besides the teacher's desk at the front and students' desks set up in groups of between four and six, and the storage cupboard at the back of the room. From my discussion with the teacher prior to the lesson, I know that this room is used by a number of teachers of varying specialities, not just mathematics and so she brings her books with her for each lesson.

Once all the students have arrived, there are 25 grade 9 girls in the class who seemed to self-select where they would sit in the room. The lesson commences with the customary welcome by the teacher in Arabic and the unified student response. She then introduces me to the students and welcomes me to the class. I share some smiles with the students who then redirect their attention back to the teacher as she explains the plan for the lesson. The intention is for the students to learn how to solve trigonometric application-type problems set in real-life contexts. The teacher asks the students to collect the textbooks from the storage at the back of the room and to turn to a specific page. She explains that students will be solving problems related to angles of elevation and depression and she physically demonstrates these by putting one arm on the horizontal and moving the other above and below to show each type of angle. The lesson progresses by the teacher modelling the solution, on the whiteboard, to a question from the textbook. The question involves calculating the distance that two people are apart on opposite sides of a tower. As she demonstrates the answer, she asks the class some questions, "how do we label the triangle?" "What trig ratio do we use?" "What is the ratio for $sin\theta$?" "What units do we need?" Students put their hands up after each question and the teacher asks one student to answer each time.

> Once the demonstration question has been explained, the teacher allocates the next question from the book for students to complete individually. While the students are working on the problem, the teacher remains at the front of the room and after approximately one minute, she stops the class and proceeds to start the explanation of the solution. This involves one student coming to the board and writing the working and the answer. I notice that many students have waited till the answer is being written on the board before starting to write the solution in their books. The next question is then allocated, and students are expected to complete it. From my seated position at the back of the room, I can clearly see the work of the group closest to me. One student is busy solving while the others have not yet started, then once she finishes, she angles her page so that the rest of the group can copy the working into their books. The same process then ensues with one student volunteering to come to the board and complete the solution. All students are working at the same pace as only one question is allocated at a time. This means that some students are finished quickly and sit with nothing to do while others wait for the answers to be modelled on the board and then quickly write it in their books.
>
> During this phase of the lesson, the teacher remains at the front of the room, either at the whiteboard or sitting at her desk. From my vantage point, I notice one group of girls where two students are doodling in their exercise books, one student has pulled the cover off the textbook and proceeds to rip it into small pieces, and the last student is passing notes to the girls at the next group of tables. None of these students are being particularly disruptive and so the teacher does not appear to notice their behaviour however, they do not seem to be engaged in the allocated task. The lesson continues with a question being allocated, one minute to solve, and then a show of hands as to who would like to demonstrate their solution. All the questions are from the textbook and involve cycling, bearings and sailing, and towers.
>
> About 20 minutes before the end of the lesson, a girl at a group nearby falls asleep. Her friend sees me watching her and smiles, then looks at the teacher to see if she has noticed. When she realises the teacher has not seen, she proceeds to play with the hair of the girl sitting on the other side of her. They continue this for the rest of the class time. Just prior to the end, it is time to wake the sleeping student and she appears very disorientated when awoken. Her friends giggle and smile. The bell then rings and the students leap up, depositing the textbooks back in the cupboard as they quickly exit the room. There is no opportunity for a plenary or closing discussion before the room is empty.

The behaviours of the students, as described in this impressionistic tale, indicated that they were not engaged in the task. The students did not appear to be listening either to the teacher or to each other and were engaged in off-task behaviour such as drawing on their books and, in one case, falling asleep. The impressionistic tales are analysed by the identification and discussion of themes in order to compare and contrast the two contexts of classrooms with teachers exemplary in the use of explorations and those who were not.

Discussion and Recommendations

With consideration of the intention of the UAE government to continue to focus on developing a knowledge-based economy with students pursuing STEM fields at university and in careers, the importance of innovative teaching and learning approaches that increase student engagement cannot be ignored. Analysis of the quantitative and qualitative data has indicated that the use of inquiry-based learning positively impacted students' perceptions of the learning environment and attitudes towards mathematics. While all the survey results showed more positive perceptions in classes with teachers exemplary in the use of explorations, the main themes from the qualitative data pertained to these particular variables: student engagement and involvement in learning; task value and real-life application and inquiry and investigation.

Student Engagement and Involvement in Learning

As stated by Dary et al. (2016, p. 5):

> Student engagement occurs when young people have invested themselves, their energy, and their commitment to the learning environment, both within and outside the classroom. They willingly put forth the required effort to find a level of personal success academically, socially, and emotionally.

Using the LEIS construct as a guide, involvement can be described as the extent that students participate in discussions, have attentive interest, enjoy their class and complete additional work. Increased student engagement and involvement were observed in classes with teachers exemplary in the use of explorations in a variety of ways, including active learning, opportunities to share information and collaborate with peers and using higher order thinking. Each of these is described further below.

Active Learning

In classes where students were taught by teachers exemplary in the use of inquiry, more opportunities for active learning were observed than for those in classes where the teacher was not. Students were able to direct the learning of content and process, which was observed in the group tasks, with students huddling together over their presentation paper discussing where each piece of information should be displayed. Observing the class, it was noted that the majority of the students were engaged in on-task behaviours. Due to the variety of tasks to be completed, there was sufficient work for everyone, but there was also an element of choice over the responsibilities each student would undertake within the group.

In contrast, students in classes with teachers who were not exemplary in the implementation of inquiry, had less opportunities to be active in the learning process. During observations it was noted that students completed tasks more slowly, including collecting books, organising workbooks and being more hesitant to commence solving problems, than their counterparts in classes with an exemplary teacher. All students completed the same tasks, and in fact, many waited till others had gained the solution to problems before copying from the student by them or from the whiteboard. Throughout the observations, very rarely were the students heard to be dialoguing about the content or mathematical processes involved in gaining answers.

Opportunities to Collaborate and Share with Peers

Second, students in classes with teachers exemplary in their use of explorations described more opportunities to collaborate with their peers and to share information with the teacher during the focus group interviews. They explained that the learning within their group was increased by sharing and working together. For example, one student stated "We like to work together to research. She say [s] something, then she say [s] something, then I say something. Then we all learn. We have much information" [Student 1]. The students valued the opportunity to share ideas as they explained that this allowed them to better understand the mathematical content they were learning.

In contrast, students taught by teachers who were not exemplary in their use of explorations described less opportunities to collaborate. During the focus group interviews the researcher asked if the students could describe a time when they worked with their classmates and shared information in a collaborative context. The students remained quiet initially. There was confusion over the meaning of collaboration, but once the students had established understanding through defining the term and then translating into Arabic, a student explained that she had worked together with her friends on a worksheet. She stated "the clever girl in our group mostly completed the questions, but we helped sometimes" [Student 7].

Use of Higher Order Thinking

Finally, students in classes with teachers exemplary in the use of explorations were observed to be more engaged and involved in their learning and experienced more opportunities to utilise higher order thinking skills. These included activities where they could think critically and justify their responses within the tasks. Within the lesson observations, it was noted that students questioned each other within their groups and provided justifications concerning their answers and the process involved. For example, the researcher overheard one student say to another "… but why do you think that? Couldn't it be this instead?" [Student 4]. The second student then justified her conclusions by stating "It has to be within this range as the rocket must be faster than the plane. This calculation must be wrong as the answer isn't sensible" [Student 5]. Conversations observed in the classroom also included discussions concerning the limitations of their model and the assumptions they had made.

Contrasting the observations in classes with teachers who were not exemplary in their use of explorations, students had less opportunities to benefit from higher order thinking skills. During the lessons, it was observed that the teachers predominantly stayed at the front of the classroom and that discussions involving individuals or small groups of students tended not to occur. There were no obvious examples of teachers questioning students on why they had made particular calculations or to justify the solutions to the problems. Instead, the focus appeared to be on completing problems from the textbooks. It should be noted that the questions were word problems involving applications; however, all required the same structure of solution and thinking: draw a diagram, label, define the appropriate trigonometric ratio, substitute and solve. Opportunities to engage in higher order thinking skills were not widely observed in these classes.

Task Value and Real-Life Application

Through the analysis of the data, the second theme that emerged indicated that students in classes with teachers exemplary in the use of explorations had more opportunities to relate their work to their real lives and expressed more value for the tasks they were required to complete. Research indicates that when students value a task, they express greater motivation and a willingness to persevere than those who do not (Eccles & Wigfield, 2002; Eccles et al., 1983; Wigfield & Cambria, 2010). The SATMS scale, task value, assesses the extent that students believe the task they are completing is worthwhile, important and useful (Velayutham et al., 2011). Analysing the qualitative data helped to better understand the quantitative results that showed scores in the task value scale were higher for students in classes with teachers exemplary in the use of explorations than those who were not. There is a relationship between the two aspects of the second theme, task value and real-life applications. Where students can see the relevance of a task to their personal lives, particularly with reference to their future careers, the students often value it more (Assor et al., 2002). During the focus group interviews and classroom observations, increased task value and opportunities for applications to real-life were noted in classes with teachers exemplary in the use of explorations. Aspects of these included: links to other subjects, in particular to science, opportunities for the application of mathematical content in their own lives and an increase in finding tasks useful and meaningful. These are explained further subsequently.

Links to Other Subjects

First, students in classes with teachers exemplary in the use of explorations were observed to have more opportunities to link their mathematics to other subjects. For example, during a lesson observation, it was noted that students were exploring how long it would take them to fly to Pluto. The links to science were evident with students carrying out research involving planets, distances and calculations involving scientific notation to find the answers to their initial questions. The task required students to think beyond the algorithm involved with calculations using scientific notation and consider the purpose of using scientific notation when numbers were too large, as in distances to planets. Students were able to understand why it is essential to utilise scientific notation in calculations of this nature. During the focus groups, students were able to elaborate on the focus of linking with science and how much they enjoyed this approach.

> I had never thought about space travel before this class. Although we were pretending that we could fly to Pluto in an aeroplane, it still makes me think about actual astronauts and their space travel. I'm keen to read and learn more about this as I find it very interesting [Student 3].

In contrast, students in classes with teachers not exemplary in the use of explorations had fewer opportunities to relate mathematics to additional subjects at school. In the lesson observations, students completed questions from a textbook and although in one example a question involving sailing directions and distances had a tenuous link to geography, this was not the norm in the other observed classes. During the focus group interviews, the students were asked to describe a time when they related the mathematics they were learning in class, to work in their other subjects. After a period of silence, one student shared an experience of solving a word problem and having difficulty understanding some of the text as the question was in English. She explained that she had used a dictionary to understand the words, so then she could solve the problem "Just like we do in English!" [Student 8]. This was the only anecdote that the students from classes with teachers who were not exemplary in the use of explorations, were able to share.

Opportunities for Application of Mathematical Content

Second, analyses of the qualitative information indicated that students in classes with teachers exemplary in the use of explorations had more opportunities to relate their mathematics learning to their own lives. For example, the classes described in the first impressionistic tale, it was observed that students used the opportunity to research distances to planets and speeds of various transport to think about their potential future careers and apply the learning to their personal lives. In one case, a student appeared to be excited to be learning about NASA and space as part of the research of her group. She was eager to tell the researcher that she wanted to be an astronaut and expressed interest in the space programmes that the UAE was participating in. When digging deeper on these topics during the focus group interviews, students were able to articulate how they felt about the work that they were learning and the way they were learning. For example, one student said:

> Most classes at school teach us in a narrow way. We memorise our work for long enough to pass the exam. In this math class, we see the work in a bigger way. How it fits into careers and research into science and the whole point of learning the topic. I find this helps me to remember better. I don't need to memorise as I'm experiencing it instead [Student 4].

In contrast, students in classes with teachers who were not exemplary in the use of explorations found less opportunities to relate the mathematical content to their own lives. In the lesson observations, students completed activities that required them to complete questions from the textbook. Many questions involved contexts related to sailing boats, but the students lived in a small community, a number of hours' drive from the sea. In other questions, the diagrams involved large towers; however, in the community in which they lived, all buildings were two storeys or less.

Usefulness of Tasks

Finally, based on my observations and interviews with students, it would appear that students in classes with teachers exemplary in the use of explorations found the tasks provided to be more valuable and meaningful than their counterparts. Task value has four components: intrinsic value (how much enjoyment the individual derives from the task); utility value (how the task relates to future goals); attainment value (the perceived importance of doing well) and cost value (what the individual has to give up to engage in a task) (Eccles & Wigfield, 2002). The first two components of task value seemed to be the most prevalent within the lesson observations, with students enjoying the work they were completing and finding the topics and style of learning important for their futures.

Students in classes with teachers exemplary in the use of explorations appeared to experience a variety of valuable opportunities from the exploration tasks they were completing. During one of the lesson observations, while the researcher was moving between groups and discussing the set task with them, one student explained that she felt learning through explorations allowed her to develop new skills (outside of the mathematical content) that she believed were beneficial for success in the unit tests, exams and future life. She stated, "Now that we are learning most of our math with explorations, I have learned to think more about why and how I am working and what the math means for life. It has helped me in questions in exams, that aren't just the normal remembering ones, that have a context" [Student 6].

In contrast, observations and interviews indicate that students in classes with teachers who were not exemplary in the use of explorations had less opportunities to find value in the work they were completing. During the lesson observations, some students chose not to complete the tasks that had been allocated to them. For example, a group of students chose to sit and write notes to one another and did not attempt to complete the questions from the textbook. In another example, the researcher observed four students, sitting at adjoining desks, who chose not to discuss the work they were completing. Instead, one student completed the calculations for the answer and then the others in the group copied this onto their pages once she had finished. During the focus group interviews, to further understand the behaviour the researcher had observed in the lessons, they asked about whether the students found the activities the teacher set, worthwhile and valuable. One student, whose response was representative of others, stated:

> Math is boring. It's always the same. I like subjects that will help me get a good job and they say that I will need math for my career, but I can't see how. It's just numbers that don't relate to anything else [Student 9].

Inquiry and Investigation

The third theme that emerged from the analysis of the qualitative data was that students in classes with teachers exemplary in the use of explorations experienced a learning environment that had greater inquiry and investigation when compared to students in classes with teachers non-exemplary in the use of explorations. Inquiry in this context involved the use of the inquiry process; that is, the development of rich questions to guide the inquiry, research to gather relevant information, synthesis of the information and demonstration of the findings (Jansen, 2011). Based on the investigation scale of the LEIS, an investigation is described as "the degree to which skills and processes of inquiry and their use in problem-solving and investigation are emphasised". Investigation, therefore, involves providing students with opportunities to explore and discover mathematics in a student-centred manner rather than a didactic teacher-led situation where memorising facts that are given is the norm. During classroom observations, in classes taught by teachers exemplary in the use of explorations, increased opportunities for inquiry-based learning and investigation were observed in a number of ways, including opportunities for students to use open-ended tasks; research from multiple sources and opportunities for students to reflect on their work and think critically. Each of these is expanded upon below.

Opportunities to Use Open-Ended Tasks

First, students in classes with teachers exemplary in the use of explorations were given more opportunities to complete open-ended tasks. Open-ended tasks are those that have multiple possible answers, allowing students insights into a range of mathematical opportunities through seeing and discussing (Sullivan et al., 2009). Such tasks have been found to be influential in supporting student opportunities for exploration, collaboration and mathematical reasoning (Kosyvas, 2016). Analysis of the data collected during the lesson observations and focus group interviews indicated that in classes with teachers exemplary in the use of explorations, lessons involved tasks with an open-ended nature which allowed students to take their group's investigation in a different direction from other groups in the class. For example, during one lesson while moving between groups of students, it was noted that each group had taken a different approach to the task that had been set. Some students had chosen to use a commercial plane as their mode of transport while others had chosen to focus on using a rocket. Similarly, different groups of students had chosen differing methods of presentation, both for their calculations and mathematical working and for the verbal and written presentations. One student explained to the researcher, "I really like that each group can go in their own direction in solving the problem. It makes it interesting at the end when we present our work. Seeing what everyone has done". [Student 10].

In contrast, students in classes with teachers who were not exemplary in the use of explorations experienced less opportunities to explore their learning through open-ended tasks. Instead, students were observed completing closed tasks where every student completed the same steps to attain the same solution. For example, in the impressionistic tale, the teacher demonstrating the answer to one of the questions and instructing the students to complete the next question from the textbook was described. In this case, the teacher gave the students time to commence work on the mathematics problem but started to set up the solution on the white board (by labelling the triangle and explaining the next steps) before they had finished. In this example, many of the students waited until the teacher started solving the problem and then wrote the answer from the board into their books. The students then went on to complete the same tasks that had fixed solutions for all.

Research from Multiple Sources

Second, students in classes with teachers exemplary in the use of explorations had opportunities to use multiple sources for research as part of their learning process. In traditional mathematics lessons, students tend to receive their information directly from the teacher, often using the textbook to support with examples of model question types. An aspect of inquiry-based learning is enabling students to be able to gather information and data from a number of differing sources. This could be from experts, within and external to the school, from books in the library and using online sources of information. For example, during one of the lesson observations, students explained that, during earlier lessons, they had been given the 'big question' and then, in groups, they had planned how they would answer it. Some of the decisions that they were required to make included how they would plan the research that they would need to complete the inquiry and how they would present their findings to the class at the end of the exploration. During a focus group interview, these students articulated the inquiry process that they were undergoing, including the variety of research they would undertake.

> We follow a process that our teacher has taught us where we start with a question that is quite wide and big and then in our groups, we narrow our question and make it more specific for what we want to investigate. It's then very important for us to decide how to gather our information. When we first started learning this way it was hard for us to be able to work out for ourselves how to answer the big question, but as we did it more, we became better at knowing what to do [Student 3].

In contrast, students in classes with teachers who were not exemplary in the use of explorations had less opportunities to research or use multiple sources for information. Inquiry-based learning classes require the teacher to be a facilitator rather than an expert lecturer. Observations indicated that, in these classes, predominantly, the teacher was the sole source of information for the students. For example, in one class, the teacher asked the students to complete a question from the textbook and then remained at the front of the classroom, initially standing and then sitting at her desk. From her viewpoint, she kept an eye on the students by her desk and when they appeared to be finished, she got up and went to the whiteboard where she requested one student to join her and asked them to solve the problem on the board for all students to see. During the focus group interviews, when discussing the use of inquiry during their mathematics lessons, these students appeared to have little understanding of the meaning of inquiry. That is, for the most part, the students generally agreed that, when they were asked to solve a problem individually after being given an example on the board, this was an inquiry. The researcher asked the students in the focus groups to explain a time during their mathematics lessons when they used inquiry to learn. One student stated, "We use inquiry in class a lot. The teacher shows us a question on the board and the steps to solve, and then we try one in our books" [Student 7]. Another student agreed, "Yes, the teacher gets us to solve the next problem by ourselves, so we have to think. Just like inquiry!" [Student 8].

Opportunities for Reflection and Critical Thinking

Third, students in classes with teachers exemplary in the use of explorations were provided with more opportunities to think critically and to reflect on their learning processes. An important goal of inquiry-based learning is to foster critical thinking in students (National Research Council, 1996). Critical thinking is defined as something which "involves the use of information, experience, and world knowledge in ways which allow students to seek alternatives, make inferences, pose questions, and solve problems, thereby signalling understanding in a variety of complex ways" (Liaw, 2007, p. 51). For example, in classes taught by teachers exemplary in the use of explorations, students were given opportunities to discuss the limitations of their research model. In one case, students were observed describing their research and findings and discussing the limitations of their model. As part of this process, the students articulated where they believed they had been successful and what parts of the process they would change for next time. As one student explained:

> The first time we did an exploration task I found it very bad. It was so hard to think in this new way. Before we just memorised the equations and did the questions in the book, but now we had to explain what we were doing and why. I am much better at it now and can even think of better ways to do it next time [Student 12].

In contrast, students who were in classes with teachers who were not exemplary in the use of explorations rarely had opportunities to think critically and to reflect on their work. Rather, these students completed the same algorithmic process for each trigonometric problem from the textbook involving the steps: draw a diagram, label sides, choose trigonometric ratio, substitute and solve. When students encountered difficulties, they waited for the teacher to write the solution on the board so that they could copy it down, or alternatively asked their friends if they could see their answer. It was not evident that any students had opportunities to think critically about the task they were completing. As part of the questioning in the discussions in the focus group interviews, students were asked if they could describe a time when they reflected on their work, either on the answers they gained or the process they underwent to attain the solution. One student explained that she likes to think about whether her answer is sensible in the context of the question. By probing further, the researcher enquired about why she does this, and she answered "Last year in our math class, our teacher told us to always think about the units and the number and whether they work. This year we don't do that, but I like to" [Student 18]. This was the only evidence of critical thinking or reflective practice that was observed during the lessons or within the focus group interviews. In general, the students in classes with teachers who were not exemplary in the use of explorations did not have opportunities for reflective or critical thinking.

Recommendations and Future Research

Considering the implications of the analysis of the quantitative and qualitative data, a number of recommendations can be made. In order to build on the strong foundation with the introduction of inquiry-based learning, continued professional development opportunities should be offered to teachers to further develop their understanding and skills in supporting students with the new pedagogy. As well as workshops, a system of mentoring where teachers exemplary in the use of inquiry could support their colleagues through model teaching, sharing of resources and coaching pre- and post-lesson observations. This would enable the expertise within the professional learning community to be shared. In addition, exemplar materials developed for an inquiry-based learning classroom, specifically generated for the UAE context with appropriate cultural considerations that allow students to relate the work to their own lives, would support implementation for the teachers and students.

Further research seeking to monitor progress with the shift in pedagogy to inquiry-based learning and the impact on the learning environment and attitudes towards mathematics should be considered. Additional research seeking to establish whether relationships exist similarly for male students, for students in lower and higher grades and in other subjects could be commenced. Another aspect of further research should consider the impact of the use of inquiry-based learning on the uptake of girls in STEM courses in higher grades at school, university and then through to career choices.

In conclusion, the positive impact on girls' perceptions of the learning environment and their attitudes towards mathematics as a result of well-taught inquiry-based learning in mathematics classes in middle schools in Abu Dhabi, helps to support the educational reform agenda and future goals of the country in seeking to promote STEM courses and careers for female students.

References

Abu Dhabi Education Council. (2013). *Cycle 2 mathematics curriculum—Teacher guidebook*. Abu Dhabi: Abu Dhabi Education Council.

Acee, T. W., Weinstein, C. E., Hoang, T. V., & Flaggs, D. A. (2018). Value reappraisal as a conceptual model for task-value interventions. *Journal of Experimental Education, 86*(1), 69–85. https://doi.org/10.1080/00220973.2017.1381830

Afari, E., Aldridge, J. M., & Fraser, B. J. (2012). Effectiveness of using games in tertiary-level mathematics classrooms. *International Journal of Science and Mathematics Education, 10*(6), 1369–1392. https://doi.org/10.1007/s10763-012-9340-5

Ahmed, W., van der Wef, G., Kuyper, H., & Minnaert, A. (2013). Emotions, self-regulated learning, and achievement in mathematics: A growth curve analysis. *Journal of Educational Psychology, 105*, 150–161. https://doi.org/10.1037/a0030160

Akaygun, S., & Adadan, E. (2021). Fostering senior primary school students' understanding of climate change in an inquiry-based learning environment. *Education 3–13, 49*(3), 330–343. https://doi.org/10.1080/03004279.2020.1854961

Aldridge, J. M., Afari, E., & Fraser, B. J. (2012a). Influence of teacher support and personal relevance on academic self-efficacy and enjoyment of mathematics lessons: A structural equation modeling approach. *Alberta Journal of Educational Research, 58*(4), 614–633. https://doi.org/10.1037/t38960-000

Aldridge, J. M., Fraser, B. J., Bell, L., & Dorman, J. (2012b). Using a new learning environment questionnaire for reflection in teacher action research. *Journal of Science Teacher Education, 23*(3), 259–290. https://doi.org/10.1007/s10972-012-9268-1

Aldridge, J. M., Fraser, B. J., & Huang, T.-C.I. (1999). Investigating classroom environments in Taiwan and Australia with multiple research methods. *Journal of Educational Research, 93*(1), 48–62. https://doi.org/10.1080/00220679909597628

Alt, D. (2018). Teachers' practices in science learning environments and their use of formative and summative assessment tasks. *Learning Environments Research, 21*(3), 387–406. https://doi.org/10.1007/2Fs10984-018-9259-z

Amaral, O., Garrison, L., & Klentschy, M. (2002). Helping English learners increase achievement through inquiry-based science instruction. *Bilingual Research Journal, 26*(2), 213–239. https://doi.org/10.1080/15235882.2002.10668709

Anaya, L., Stafford, F., & Zamarro, G. (2021). Gender gaps in math performance, perceived mathematical ability and college STEM education: The role of parental occupation. *Education Economics*. https://doi.org/10.1080/09645292.2021.1974344

Anderson, A., Hamilton, R. J., & Hattie, J. A. (2004). Classroom climate and motivated behaviour in secondary schools. *Learning Environments Research, 7*, 211–225. https://doi.org/10.1007/s10984-004-3292-9

Assor, A., Kaplan, H., & Roth, G. (2002). Choice is good, but relevance is excellent: Autonomy-enhancing and suppressing teacher behaviours in predicting student's engagement in school work. *British Journal of Educational Psychology, 72*, 261–278. https://doi.org/10.1348/000709902158883

Baeten, M., Dochy, F., & Struyven, K. (2013). The effects of different learning environments on students' motivation for learning and their achievement. *British Journal of Educational Psychology, 83*(3), 484–501. https://doi.org/10.1111/j.2044-8279.2012.02076.x

Bain, J. (2010). Integrating student voice: Assessment for empowerment. *Practitioner Research in Higher Education, 4*(1), 14–29. http://dx.doi.org/1755-1382

Bandura, A. (1977). Self-efficacy: Toward a unifying theory of behavioral change. *Psychological Review, 84*(2), 191–215. https://doi.org/10.1037/0033-295X.84.2.191

Breckler, S. J. (1984). Empirical validation of affect, behavior, and cognition as distinct components of attitude. *Journal of Personality and Social Psychology, 47*(6), 1191–1205. https://doi.org/10.1037/0022-3514.47.6.1191

Bruder, R., & Prescott, A. (2013). Research evidence on the benefits of IBL. *ZDM the International Journal on Mathematics Education, 45*(6), 811–822. https://doi.org/10.1007/s11858-013-0542-2

Camenzuli, J., & Buhagiar, M. A. (2014). Using inquiry-based learning to support the mathematical learning of students with SEBD. *International Journal of Emotional Education, 6*(2), 69–85.

Chapman, O., & Heater, B. (2010). Understanding change through a high school mathematics teacher's journey to inquiry-based teaching. *Journal of Mathematics Teacher Education, 13*(6), 445–458. https://doi.org/10.1007/s10857-010-9164-6

Chionh, Y. H., & Fraser, B. J. (2009). Classroom environment, achievement, attitudes and self-esteem in geography and mathematics in Singapore. *International Research in Geographical and Environmental Education, 18*(1), 29–44. https://doi.org/10.1080/10382040802591530

Choi, N. (2005). Self-efficacy and self-concept as predictors of college students' academic performance. *Psychology in the Schools, 42*(2), 197–205. https://doi.org/10.1002/pits.20048
Clarke, V., & Braun, V. (2013). Teaching thematic analysis: Overcoming challenges and developing strategies for effective learning. *The Psychologist, 26*(2), 120–123.
Cohen, J. (1988). *Statistical power analysis for the behavioral sciences*. Lawrence Erlbaum Associates.
Creswell, J. W. (2008). *Educational research*. Pearson Prentice Hall.
Creswell, J. W., & Plano Clark, V. L. (2017). *Designing and conducting mixed methods research* (3rd ed.). SAGE.
Dary, T., Pickeral, T., Shumer, R., & Williams, A. (2016). Weaving student engagement into the core practices of schools: A national dropout prevention center/network position paper. Retrieved from Clemson, SC: http://dropoutprevention.org/resources/major-research-reports/studentengagement/student-engagement-2016-09.pdf
Dewey, J. (1920, 1948). *Reconstruction in philosophy*. Beacon Press.
Dorman, J. P. (2008). Use of multitrait-multimethod modelling to validate actual and preferred forms of the What Is Happening In this Class? (WIHIC) questionnaire. *Learning Environments Research, 11*(3), 179–193. https://doi.org/10.1007/s10984-008-9043-6
Eccles, J. S., Adler, T. F., Futterman, R., Goff, S. B., Kaczala, C. M., & Meece, J. L. (1983). Expectancies, values, and academic behavior. In J. T. Spencer (Ed.), *Achievement and achievement motivation* (pp. 75–146). W. H. Freeman.
Eccles, J. S., & Wigfield, A. (2002). Motivational beliefs, values, and goals. *Annual Review of Psychology, 53*, 109–132. https://doi.org/10.1146/annurev.psych.53.100901.135153
Ellis, K. (2004). The impact of perceived teacher confirmation on receiver apprehension, motivation, and learning. *Communication Education, 53*(1), 1–20. https://doi.org/10.1080/0363452032000135742
Field, A. (2016). *Discovering statistics using IBM SPSS statistics* (5th ed.). SAGE.
Fielding-Wells, J., O'Brien, M., & Makar, K. (2017). Using expectancy-value theory to explore aspects of motivation and engagement in inquiry-based learning in primary mathematics. *Mathematics Education Research Journal, 29*(2), 237–254. https://doi.org/10.1007/s13394-017-0201-y
Firestone, W. A. (1987). Meaning in method: The rhetoric of quantitative and qualitative research. *Educational Researcher, 16*(7), 16–21. https://doi.org/10.3102/0013189X016007016
Fraser, B. J. (2012). Classroom learning environments: Retrospect, context and prospect. In B. J. Fraser, K. G. Tobin, & C. J. McRobbie (Eds.), *The second international handbook of science education* (pp. 1191–1239). Springer.
Fraser, B. J. (2015). Classroom learning environments. In R. Gunstone (Ed.), *Encyclopedia of science education* (pp. 154–157). Springer. https://doi.org/10.1007/978-94-007-2150-0_186
Fraser, B. J. (2018). Milestones in the evolution of the learning environments field over the past three decades In D.B. Zandvliet & B.J. Fraser (Eds.), *Thirty years of learning environments: Looking back and looking forward* (pp. 1–19). Brill Sense. https://doi.org/10.1163/9789004387720
Gaad, E., Arif, M., & Scott, F. (2006). Systems analysis of the UAE education system. *International Journal of Educational Management, 20*(4), 291–303. https://doi.org/10.1108/09513540610665405
Gaspard, H., Dicke, A.-L., Flunger, B., Schreier, B., Häfner, I., Trautwein, U., & Nagengast, B. (2015). More value through greater differentiation: Gender differences in value beliefs about math. *Journal of Educational Psychology, 107*, 663–677. https://doi.org/10.1037/edu0000003
Gholam, A. (2019). Inquiry-based learning: Student teachers' challenges and perceptions. *Journal of Inquiry & Action in Education, 10*(2), 112–133. https://digitalcommons.buffalostate.edu/jiae/vol10/iss2/6
González-Pérez, S., Mateos de Cabo, R., & Sáinz, M. (2020). Girls in STEM: Is it a female role-model thing? *Frontiers in Psychology, 11*(2204). https://doi.org/10.3389/fpsyg.2020.02204.
Goetz, T., Nett, U. E., Martiny, S. E., Hall, N. C., Pekrun, R., Dettmers, S., & Trautwein, U. (2012). Students' emotions during homework: Structures, self-concept antecedents, and achievement

outcomes. *Learning and Individual Differences, 22*(2), 225–234. https://doi.org/10.1016/j.lindif.2011.04.006

Heindl, M., & Nader, M. (2018). Digital technologies in a design and technology lesson and their influence on a learner's situationally perceived value of a task when engaged in inquiry-based learning. *Journal of Computers in Mathematics and Science Teaching, 37*(3), 239–263.

Hidi, S., & Renninger, K. A. (2006). The four-phase model of interest development. *Educational Psychologist, 41*, 111–127. https://doi.org/10.1207/s15326985ep4102_4

James, W. (1907, 1995). *Pragmatism*. Dover.

Jansen, B. A. (2011). Inquiry unpacked: An introduction to inquiry-based learning. *Library Media Connection, 29*(5), 10–12.

Kang, J., & Keinonen, T. (2017). The effect of inquiry-based learning experiences on adolescents' science-related career aspiration in the Finnish context. *International Journal of Science Education, 39*(12), 1669–1689. https://doi.org/10.1080/09500693.2017.1350790

Kogan, M., & Laursen, S. L. (2014). Assessing long-term effects of inquiry-based learning: A case study from college mathematics. *Innovative Higher Education, 39*, 183–199. https://doi.org/10.1007/s10755-013-9269-9

Koh, N. K., & Fraser, B. J. (2014). Learning environment associated with use of mixed mode delivery model among secondary business studies students in Singapore. *Learning Environments Research, 17*(2), 157–171. https://doi.org/10.1007/2Fs10984-013-9139-5

Kosyvas, G. (2016). Levels of arithmetic reasoning in solving an open-ended problem. *International Journal of Mathematical Education in Science and Technology, 47*(3), 356–372. https://doi.org/10.1080/0020739x.2015.1072880

Laine, E., Veermans, M., Lahti, A., & Veermans, K. (2017). Generation of student interest in an inquiry-based mobile learning environment. *Frontline Learning Research, 5*(4), 42–60. https://doi.org/10.14786/flr.v5i4.306

Lauermann, F., Tsai, Y., & Eccles, J. S. (2017). Math-related career aspirations and choices within Eccles et al.'s expectancy-value theory of achievement-related behaviors. *Developmental Psychology, 53*(8), 1540–1559. https://doi.org/10.1037/dev0000367

Lazarides, R., & Lauermann, F. (2019). Gendered paths into STEM-related and language-related careers: Girls' and boys' motivational beliefs and career plans in math and lanuage arts. *Frontiers in Psychology, 10*(1243). https://doi.org/10.3389/fpsyg.2019.01243

Lazarides, R., & Watt, H. M. G. (2015). Student-perceived mathematics teacher beliefs, math classroom learning environments and gendered math career intentions. *Contemporary Educational Psychology, 41*, 51–61. https://doi.org/10.1016/j.cedpsych.2014.11.005

Leikin, R., & Rota, S. (2006). Learning through teaching: A case study on the development of a mathematics teacher's proficiency in managing an inquiry-based classroom. *Mathematics Education Research Journal, 18*(3), 44–68. https://doi.org/10.1007/BF03217442

Liaw, M. L. (2007). Content-based reading and writing for critical thinking skills in an EFL context. *English Teaching and Learning, 31*, 45–87. https://doi.org/10.6330/ETL.2007.31.2.02

Makar, K. (2007). Connection levers: Supports for building teachers' confidence and commitment to teach mathematics and statistics through inquiry. *Mathematics Teacher Education and Development, 8*, 48–73.

Marshall, J. C., & Horton, R. M. (2011). The relationship of teacher-facilitated, inquiry-based instruction to student higher-order thinking. *School Science and Mathematics, 111*(3), 93–101. https://doi.org/10.1111/j.1949-8594.2010.00066.x

Marshall, J. C., Smart, J., & Horton, R. M. (2010). The design and validation of EQUIP: An instrument to assess inquiry-based instruction. *International Journal of Science and Mathematics Education, 8*(2), 299–321. https://doi.org/10.1007/s10763-009-9174-y

Merton, R. K. (1975). Thematic analysis in science: Notes on Holton's concept. *Science as Culture, 188*(4186), 335–338. https://doi.org/10.1126/science.188.4186.335

Morgan, D. L. (1988). *Focus groups as qualitative research*. SAGE.

National Research Council. (1996). *National science education standards*. National Academy Press.

Neale, D. (1969). The role of attitudes in learning mathematics. *The Arithmetic Teacher, 16*(8), 631–641.

Nebesniak, A. L., & Heaton, R. M. (2010). Student confidence and student involvement. *Mathematics Teaching in the Middle School, 16*(2), 96–103.

OECD. (2020). Girls' and boys' performance in PISA, in PISA 2018 results (Volume II): Where all students can succeed. *OECD Publishing*. https://doi.org/10.1787/f56f8c26-en

Oliver, R. (2007). Exploring an inquiry-based learning approach with first-year students in a large undergraduate class. *Innovations in Education and Teaching International, 44*(1), 3–15. https://doi.org/10.1080/14703290601090317

Ozkal, K., Tekkaya, C., Cakiroglu, J., & Sungur, S. (2009). A conceptual model of relationships among constructivist learning environment perceptions, epistemological beliefs, and learning approaches. *Learning and Individual Differences, 19*(1), 71–79. https://doi.org/10.1016/j.lindif.2008.05.005

Parlett, M., & Hamilton, D. (1976). Evaluation as illumination. In D. Tawney (Ed.), *Curriculum evaluation today: Trends and implications* (pp. 84–101). Macmillan.

Partin, M. L., & Haney, J. J. (2012). The CLEM model: Path analysis of the mediating effects of attitudes and motivational beliefs on the relationship between perceived learning environment and course performance in an undergraduate non-major biology course. *Learning Environments Research, 15*(1), 103–123. https://doi.org/10.1007/2Fs10984-012-9102-x

Peirce, C. S. (1878). How to make our ideas clear. *Popular Science Monthly, 12*, 286–302.

Pekrun, R., Goetz, T., Titz, W., & Perry, R. P. (2002). Positive emotions in education. In E. Frydenberg (Ed.), *Beyond coping: Meeting goals, visions, and challenges* (pp. 149–173). Oxford University Press.

Ponterotto, J. G. (2005). Qualitative research in counseling psychology: A primer on research paradigms and philosophy of science. *Journal of Counseling Psychology, 52*(2), 126–136. https://doi.org/10.1037/0022-0167.52.2.126

Priniski, S. J., Hecht, C. A., & Harackiewicz, J. M. (2018). Making learning personally meaningful: A new framework for relevance research. *Journal of Experimental Education, 86*(1), 11–29. https://doi.org/10.1080/2F00220973.2017.1380589

Riegle-Crumb, C., Morton, K., Nguyen, U., & Dasgupta, N. (2019). Inquiry-based instruction in science and mathematics in middle school classrooms: Examining its association with students' attitudes by gender and race/ethnicity. *AERA Open, 5*(3), 1–17. https://doi.org/10.1177/2332858419867653

Robinson, J. M. (2020). *An investigation of the impact of inquiry-based learning in mathematics on the students' perceptions of the learning environment and attitudes towards mathematics in some Abu Dhabi schools* Curtin University]. https://espace.curtin.edu.au/handle/20.500.11937/81032

Robinson, J. M., & Aldridge, J. M. (2022). Environment-attitude relationships: Girls in inquiry-based learning classrooms in the UAE. *Learning Environments Research, 25*, 619-640. https://doi.org/10.1007/s10984-022-09409-x

Sawilowsky, S. (2009). New effect size rules of thumb. *Journal of Modern Applied Statistical Methods, 8*(2), 467–474. https://doi.org/10.22237/jmasm/1257035100

Simonson, M., & Maushak, N. (2001). Instructional technology and attitude change. In D.H. Jonassen (Ed.), *Handbook of research for educational communications and technology*. Simon & Schuster Macmillan.

Singer, F. M., & Moscovici, H. (2008). Teaching and learning cycles in a constructivist approach to instruction. *Teaching and Teacher Education: An International Journal of Research and Studies, 24*(6), 1613–1634. https://doi.org/10.1016/j.tate.2007.12.002

Spinner, H., & Fraser, B. J. (2005). Evaluation of an innovative mathematics program in terms of classroom environment, student attitudes, and conceptual development. *International Journal of Science and Mathematics Education, 3*(2), 267–293. https://doi.org/10.1007/s10763-004-6531-8

Staples, M. (2007). Supporting whole-class collaborative inquiry in a secondary mathematics classroom. *Cognition and Instruction, 25*(2), 161–217. https://doi.org/10.1080/0737000070130 1125

Sullivan, P., Griffioen, M., Gray, H., & Powers, C. (2009). Exploring open-ended tasks as teacher learning. *Australian Primary Mathematics Classroom, 14*(2), 4–9.

Sultan, W. H., Woods, P. C., & Ah-Choo, K. (2011). A constructivist approach for digital learning: Malaysian schools case study. *Journal of Educational Technology & Society, 14*(4), 149–163.

Supovitz, J. A., Mayer, D. P., & Kahle, J. B. (2000). Promoting inquiry-based instructional practice: The longitudinal impact of professional development in the context of systemic reform. *Educational Policy, 14*, 331–356. https://doi.org/10.1177/0895904800014003001

Taylor, P. C., Fraser, B. J., & Fisher, D. L. (1997). Monitoring constructivist classroom learning environments. *International Journal of Educational Research, 27*, 293–302. https://doi.org/10.1016/S0883-0355(97)90011-2

Thompson, B. (2001). Significance, effect sizes, stepwise methods, and other issues: Strong arguments move the field. *Journal of Experimental Education, 70*, 80–93. https://doi.org/10.1080/00220970109599499

Van Maanen, J. (1982). *Fieldwork on the beat*. SAGE.

Vandecandelaere, M., Speybroeck, S., Vanlaar, G., De Fraine, B., & Van Damme, J. (2012). Learning environment and students' mathematics attitude. *Studies in Educational Evaluation, 38*(3), 107–120. https://doi.org/10.1016/j.stueduc.2012.09.001

Velayutham, S., Aldridge, J. M., & Fraser, B. J. (2011). Development and validation of an instrument to measure students' motivation and self-regulation in science learning. *International Journal of Science Education, 33*(15), 2159–2179. https://doi.org/10.1080/09500693.2010.541529

Wang, M. T., & Holcombe, R. (2010). Adolescents' perceptions of school environment, engagement, and academic achievement in middle school. *American Educational Research Journal, 47*(3), 633–662. https://doi.org/10.3102/2F0002831209361209

Weaver, K. (2018). Pragmatic paradigm. In B. Frey, B. (Ed.), *The SAGE encyclopedia of educational research, measurement, and evaluation* (pp. 1287–1288). SAGE Publications, Inc. https://doi.org/10.4135/9781506326139.n534

Wigfield, A., & Cambria, J. (2010). Students' achievement values, goal orientations, and interest: Definitions, development, and relations to achievement outcomes. *Developmental Review, 30*(1), 1–35. https://doi.org/10.1016/j.dr.2009.12.001

Wolf, S. J., & Fraser, B. J. (2008). Learning environment, attitudes and achievement among middle-school science students using inquiry-based laboratory activities. *Research in Science Education, 38*(3), 321–341. https://doi.org/10.1007/s11165-007-9052-y

Wu, H.-K., & Hsieh, C.-E. (2006). Developing sixth grader's inquiry skills to construct explanations in inquiry-based learning environments. *International Journal of Science Education, 28*(11), 1289–1313. https://doi.org/10.1080/09500690600621035

Jennifer Robinson is a mathematics educator with experience in New Zealand and the United Arab Emirates. She is currently affiliated with Emirates College for Advanced Education (ECAE) in Abu Dhabi. Her research areas include learning environments, inquiry-based learning and attitudes towards mathematics. Recent publication: Robinson, J. M., & Aldridge, J. M. (2022). Environment-attitude relationships: Girls in inquiry-based learning classrooms in the UAE. *Learning Environments Research, 25*, 619–640.

Jill Aldridge is a professor at the School of Education at Curtin University in Perth, Australia. Her research interests include school improvement and system-wide reform. In particular, her research focuses on the school climate (norms, attitudes and values) and learning environment (contexts in which learning takes place) and how these can be co-constructed with young people (especially those who are most vulnerable) to ensure agency inclusivity and learning engagement. Jill is the regional editor of Springer's *Learning Environments Research: An International Journal.* She has published two books (Springer) and 62 journal articles, of which 70% were published in journals ranked by Scimago as Q1 or Q2.

Chapter 5
Mathematics Anxiety in Females—Breaking the Cycle

Melissa McMinn

Abstract The consequences of being anxious towards mathematics can be broad and long-lasting. They include the avoidance of mathematics, the limitation in selecting higher education courses and careers and negative feelings of guilt and shame. Several causes for mathematics anxiety have been reported with past educational experiences, and particularly primary school teachers, taking a sizeable amount of blame. As mathematics anxiety has been described as a wide-spread, detrimental emotion in the classroom, it is pertinent for primary school teachers to be confident in mathematics and well-prepared to be effective teachers of the subject. However, high incidences of mathematics anxiety have been repeatedly reported among in-service and pre-service teachers, and negative correlations found between mathematics anxiety and effectiveness when teaching mathematics. In particular, mathematics anxious female teachers have been found to influence girls' gender-related beliefs about who is good at mathematics, which in turn negatively affects girls' mathematics achievement. Given that females make up the majority of the primary school teaching profession in the United Arab Emirates, the context for this study, this is of concern. This chapter looks at the history of mathematics anxiety, and how it is defined and measured. The causes and consequences of mathematics anxiety, and the mathematics anxiety of UAE national pre-service teachers are discussed, and the perpetual cycle of anxiety which must be broken if we want more females in mathematics-related professions. Recommendations for breaking the cycle are made in this chapter.

Keywords Mathematics anxiety · Pre-service teachers · Teacher education · Anxiety causes

M. McMinn (✉)
Open Polytechnic, Lower Hutt, New Zealand
e-mail: Melissa.McMinn@openpolytechnic.ac.nz

M. Dickson et al. (eds.), *Gender in STEM Education in the Arab Gulf Countries*,
https://doi.org/10.1007/978-981-19-9135-6_5

Introduction

In 1954, Gough coined the term 'mathemaphobia', a term she declared unnecessary to define due to its self-explanatory nature. Gough claimed it was 'mathemaphobia' that was the root cause of many failures in mathematics classes, and that it is as prevalent as the common cold—the symptoms of which are often unnoticed until in its chronic stages. Not long after, a study by Dreger and Aiken found that many people report that they are 'emotionally disturbed in the presence of mathematics' (1957, p. 344), and labelled the syndrome 'number anxiety'. In their study of 704 students in a basic university mathematics class, they found that number anxiety appeared to be a factor separate from general anxiety, that is, an anxiety specific to dealing with numbers.

Since the 1950s, a slew of research into mathematics anxiety has been undertaken; however people have been expressing mathematics anxiety for centuries: the locution "Multiplication is vexation … and practice drives me mad" goes back at least to the sixteenth century (Dowker et al., 2016). In 1972, research demonstrated that mathematics anxiety exists among many people who do not normally suffer from any other tensions (Richardson & Suinn), and research to date is consistently finding mathematics anxiety to be a wide-spread and detrimental emotion in the classroom (see, for example, Goetz & Hall, 2013; Khasawneh et al., 2021). Mathematics anxiety has been found to affect the ways in which teachers teach in the classroom, the amount of time spent on mathematics, the commitment to new reform pedagogy, and retention in the profession (Gresham, 2018; Hemmings, 2015; Nurlu, 2015; Peker & Ertekin, 2011).

As a teacher educator, I have had many experiences of pre-service teachers regularly requesting to be assigned to lower grade level classes during practicum placements. The consistent reasoning is the fear of mathematics at the higher grade levels or the perceived inability to be able to teach it. 'I hate maths!', 'I can't do maths', 'I am not a maths person' and 'I am scared of the maths in grades four and five!' are all regular sentiments espoused by the UAE national pre-service students I work with.

This chapter investigates possible causes and consequences of mathematics anxiety, and the mathematics anxiety of UAE national pre-service teachers specifically. Firstly, some background and context on anxiety and teacher education in the United Arab Emirates (UAE) is provided.

Background and Context

To understand the milieu this chapter discusses, teacher education in the UAE is briefly described. The section goes on to review the different definitions and types of anxiety, from which a working definition of mathematics anxiety is shared and mathematics anxiety of pre-service teachers examined.

Teacher Education in the United Arab Emirates

The Abu Dhabi Economic Vision Report 2030 aims to have 90% of employees in the education sector being UAE nationals by 2030 (The Abu Dhabi Government, 2008). Four public higher educational institutes were approved by the national accreditation body to offer four-year Bachelor of Education (B.Ed.) programmes. These differed in educational specialisation, both in age-range of teaching levels and subject areas, with only two offering generalist primary teaching programmes (Mathematics, English and Science education). The programmes at both institutes were offered free exclusively to UAE national students and taught through the medium of English. For one institute, this was offered for females only, while in the other, both male and females were eligible, however male enrolment numbers were very low (<5%). Although the 'New School Model' was launched in Grades 1–3 in 2010, and a subsequent grade level each year thereafter as part of an ambitious educational reform programme in Abu Dhabi public schools, only very recently are the student beneficiaries of this reform enrolling in higher education. Reform programmes in other emirates have begun much more recently.

One higher education institute, with a sole campus in the UAE capital, Abu Dhabi, offering the primary teaching programme had dedicated classrooms and specialist teachers for the mathematics courses within the B.Ed. programme. This programme consisted of three mathematics content courses, taken over the first three semesters of the programme, designed to improve the pre-service teachers' mathematics skills and concepts. Following on were three methodology courses, taken over the subsequent three semesters of the programme and aimed at teaching mathematics at the primary school level. The second higher education institute with campuses in five locations across the UAE did not have dedicated classrooms, and any of the Education faculty could be assigned to teach the mathematics courses that constituted part of the B.Ed. programme. This programme included one mathematics content course, and one methodology course, taken concurrently in the fifth semester of the programme. Students at the second higher education institute were also required to take a mathematics course as part of the general studies requirement, taught by a member of the general studies faculty.

The main responsibility of these higher educational institutes was to prepare future UAE national teachers to teach Mathematics, Science and English in the UAE's public schools (grades 1–5) through the medium of Arabic or English, depending on the emirate in which they taught. In Abu Dhabi, a new outcomes-based curriculum, and a move towards student-centred teaching methods (New School Model) required teacher education programmes in Abu Dhabi to adapt in order to address the challenges introduced by the new curricula and to prepare teacher graduates with the skills necessary for such a change.

Definition of Anxiety

Anxiety is one of the most intensely researched constructs in the psychology field and, as such, various definitions have been espoused. Typically, general anxiety is defined with reference to a variety of unpleasant emotional reactions such as uncertainty, stress, and helplessness to real or imagined situations that are perceived as threatening or potentially dangerous. Byrd (1982, p. 5) explains that "[A]nxiety is an extremely complex phenomenon with manifestations in various areas including the affective, behavioural, and physiological. It is subjective and experiential, something people feel", and can interfere with cognitive functioning.

Anxiety can be organised into two discrete types: trait anxiety and state anxiety. Trait anxiety is defined as relatively constant individual differences in the inclination towards anxiety (Baloğlu, 1999), or the relatively consistent tendency to view situations as threatening and to react to them as such (Leso, 1992). State anxiety is a transitory emotional condition (Baloğlu, 1999) that relates directly to a stressful situation during a finite period (Leso, 1992).

Anxiety and fear are terms that are often used interchangeably (see, for example, Ashcraft, 2002; Chang & Beilock, 2016; Ma, 1999; Okur & Bahar, 2010; Ramirez et al., 2016); however, the two concepts can be distinguished by the type of threat and the reaction to that threat. With fear, the threat is known, and the reaction is physical—flight or fight—whereas with anxiety the threat is unknown, and the psychological effects of uncertainty and helplessness follow (Byrd, 1982).

Definition of Mathematics Anxiety

Many attempts have been made to define mathematics anxiety (see, for example, Ashcraft, 2002; McMinn, 2019; Richardson & Suinn, 1972; Tobias, 1980) and develop ways to diagnose and measure it; impelled by the view that the construct compromises both achievement and participation in mathematics (see, for example, Alexander & Martray, 1989; Fennema & Sherman, 1976; Gresham, 2018; Hembree, 1990; Herts & Beilock, 2017; Novak & Tassell, 2017; Plake & Parker, 1982; Ramirez et al., 2016; Richardson & Suinn, 1972; Sandman, 1980). Mathematics anxiety meets all the diagnostic criteria for a genuine phobia (Faust, 1992). Drawing on Gough's (1954) term 'mathemaphobia', Lazarus (1974) coined the term 'mathophobia', to describe an irrational, obstructive fear of mathematics. More recently, Gresham (2008) describes mathematics anxiety as a debilitating state of mind which can lead to mathematics phobia. However, Byrd (1982) warned that, while there is still disagreement, phobia, fear and anxiety are generally deemed to be distinct constructs. Similarly, the terms anxiety and attitudes have been used both synonymously and separately. Byrd (1982) makes the distinction, claiming that attitudes are primarily cognitive in nature while anxiety is primarily emotional in nature.

Despite the work that has gone into defining anxiety, mathematics anxiety has been described as difficult to define (see, for example, Baloğlu, 1999; Hembree, 1990; Rounds & Hendel, 1980; Wood, 1988) and many definitions that have been put forth have been critiqued for being too narrow (see, for example, Byrd, 1982; Wood, 1988). Possibly the most quoted definition is that of Richardson and Suinn (1972, p. 551): "Mathematics anxiety involves feelings of tension and anxiety that interfere with the manipulation of numbers and the solving of mathematical problems in a wide variety of ordinary life and academic situations". This definition is more concerned with the effect of mathematics anxiety on mathematical performance, while other definitions focus on the emotional effect on the individual. For example, "feelings of anxiety, dread, nervousness and associated bodily symptoms related to doing mathematics" (Fennema & Sherman, 1976, p. 326), or "an emotion that blocks a person's reasoning ability when confronted with a mathematical situation" (Spicer, 2004, p. 1). Byrd (1982) points out that definitions of mathematics anxiety may be limited in terms of the mathematics they refer to, often relating to arithmetic and/or problem-solving only.

In the numerous attempts to define anxiety, highly emotive terms have been used to illustrate a 'sophisticated and multi-dimensional phenomenon' (Bekdemir, 2010), including helplessness (Hunt, 1985), tension (Richardson & Suinn, 1972), apprehension (Ramirez et al., 2016), fear and angriness (Fennema & Sherman, 1976), panic (Hunt, 1985; Miller & Mitchell, 1994) and paralyzed (Morris, 1981). Several of these authors have asserted that these symptoms were physically visible (see, for example, Bekdemir, 2010; Fennema & Sherman, 1976). Perhaps with that in mind, Tobias (1980) likened mathematics anxiety to an unsurmountable feeling of sudden death.

A further complication to the development of a clear definition of mathematics anxiety is that the apprehension felt in situations as diverse as avoiding the small amount of mathematics required to check an invoice and wrestling with a complex problem on a pre-graduate mathematics examination, are often equally claimed to be a result of mathematics anxiety (Ashcraft & Moore, 2009; Wood, 1988). These, in fact, may be the result of significantly different issues:

> It becomes a subtle but important question to decide whether mathematics anxiety describes, or is a function of, some peculiar feature of the discipline of mathematics itself that produces a specific kind of anxiety that interferes with people's ability to perform mathematical tasks, or rather, whether doing mathematics produces anxiety of a more general nature (Wood, 1988, pp. 8–9).

If the former view is accepted, the focus for eliminating anxiety should be on those aspects that are intrinsic to mathematics that cause it to evoke anxiety. However, if it is the latter, then the attention should be on the societal, educational or environmental factors that cause a perception that mathematics is anxiety-producing (Wood, 1988).

Furthermore, if general anxiety can be classified as either state or trait, so too can mathematics anxiety. However, the literature is divided as to which type of anxiety, state or trait, mathematics anxiety belongs. Mathematics anxiety has been claimed to be a form of state anxiety as it occurs in mathematically related environments

(Atkinson, 1988; Baloğlu, 1999; Brady & Bowd, 2005; Byrd, 1982). State mathematics anxiety has a direct impact on performance (Buckley et al., 2016), as the cognitive burden caused by state mathematics anxiety can disrupt performance on mathematics tasks (Ashcraft & Kirk, 2001; Devine et al., 2012).

However, according to Byrd, mathematics situations can be so prevalent in modern times, that "the individual who reacts to many of them with state anxiety could be considered to have mathematics anxiety that is trait-like" (1982, p. 48). Betz's (1978) study into the mathematics anxiety of college students, found moderately strong relationships between mathematics anxiety and trait anxiety, and the Jenßen et al. (2015) study of pre-service pre-school teachers assessed mathematics anxiety to be a trait. Moreover, mathematics anxiety has been found to be stable and consistent over time and therefore can be generalised across various situations (Beilock & Maloney, 2015; Liebert & Liebert, 1998). This is supported by Klieme et al., (2008, p. 5) who found that mathematics anxiety "cannot be attributed to situational factors but rather to stable personality characteristics", suggesting that mathematics anxiety is a form of trait anxiety.

Some researchers have found mathematics anxiety to be both trait and state, depending on the individual's academic self-concept (Roos et al., 2015). Buckley et al. (2016) argued that mathematics anxiety can be both state and trait with each having different outcomes; state mathematics anxiety can negatively affect performance on mathematics tasks, while trait mathematics anxiety functions like an attitude, causing sufferers to avoid mathematics-related courses, careers and opportunities. This would suggest that individuals may suffer from either state or trait mathematics anxiety and that the differentiating factor would be the response individuals had to the mathematics anxiety: freezing up in the moment or irrationally dreading an upcoming mathematics event, or a long-term avoidance of mathematics.

Mathematics anxiety is often described in terms of its debilitating effects, physically visible symptoms and as both state and trait anxiety. For the purpose of this chapter, mathematics anxiety is defined as any negative behavioural, attitudinal or emotional reaction that impedes performance in any situation where the individual is learning, doing or being evaluated in mathematics.

Mathematics Anxiety and Test Anxiety

Many researchers have regarded mathematics anxiety and test anxiety to be highly related, yet not interchangeable (see, for example, Devine et al., 2012; Hembree, 1990), some claiming mathematics anxiety to be a form of subject-specific test anxiety (see, for example, Brush, 1981). As such, test anxiety warrants a mention in this chapter, particularly in relation to triggers and consequences, and causality. Test anxiety has been considered in terms of two models: the interference model and the deficits model.

Early researchers supported an interference model of test anxiety (see, for example, Liebert & Morris, 1967; Wine, 1971), which explains reduced performance by identifying factors (e.g. emotionality and worry) that disrupt the process of information recall during testing situations (von der Embse et al., 2018). The interference model was largely based on self-reports from students that they knew the information, but it escaped them during the exam because they became anxious (Covington & Omelich, 1987). This model assumes that students have previously learned and processed information correctly, but engage in "worrisome cognitions and task-irrelevant thoughts" (Sommer & Arendasy, 2016) during evaluation, which diverts attention away from the task and decreases their chance to retrieve the relevant information from long-term memory.

Shortly after the conceptualisation of the interference model, other theorists proposed a deficits model of test anxiety (see, for example, Culler & Holahan, 1980; Tobias, 1985; Wittmaier, 1972). According to the deficits model, high test-anxious students have difficulties in encoding and organising the material in the learning stages, either due to ineffective study habits (Birenbaum, 2007) or because the processing demands of the task are too high relative to their cognitive capacity (Tobias, 1985). This results in poor knowledge of the relevant material, meaning students have little to retrieve during the test (Birenbaum, 2007; Covington & Omelich, 1987). Consequently, test anxiety is an emotional reaction resulting from the increasing awareness of being inadequately prepared for the test (Birenbaum, 2007; Sommer & Arendasy, 2016; Tobias, 1985). As it is posited that poor performance is a direct result of a deficiency in mastery (rather than a result of anxiety), the deficit model hypothesizes that test anxiety has no causal effect on test performance (Birenbaum, 2007; Sommer & Arendasy, 2016).

Some models of test anxiety incorporate both the interference model and the deficits model (e.g. the transaction model, Spielberger & Vagg, 1995), or purport that both co-exist (the integrated model, Naveh-Benjamin et al., 1987). However, a meta-analysis of 562 test anxiety studies concluded that test anxiety was negatively correlated to performance, supporting the interference rather than the deficits test anxiety model (Hembree, 1988). While there have been significant changes in the theoretical understanding and measurement of test anxiety over the past three decades, a meta-analytic review of 238 studies from 1988 to the present by von der Embse et al. (2018) resulted in largely consistent findings with those of Hembree, suggesting that anxiety indeed interferes with performance in testing situations.

While mathematics anxiety relates to anxiety pertaining to a variety of mathematics experiences, including anxiety for mathematics in non-testing situations, it seems mathematics anxiety includes a component of text anxiety regarding mathematics evaluation which should not be overlooked. In their study with pre-service primary teachers, Uusimaki and Nason (2004, p. 374) found; "situations which caused most anxiety for the participants included communicating one's mathematical knowledge, whether in a test situation or in the teaching of mathematics such as that required on practicum".

Mathematics anxiety has been found to correlate with test anxiety more strongly than other anxiety measures (i.e. general, state and trait anxieties, Ashcraft & Moore,

2009). While Wood (1988) alleges that the possibility of separating the ideas of general anxiety, mathematics anxiety and test anxiety in a meaningful way is unclear, others have argued that mathematics anxiety may be a psychologically different construct from test anxiety (e.g. Richardson & Woolfolk, 1980). Even when a significant inter-correlation between mathematics and test anxiety measures is found, researchers make the point that the constructs are not interchangeable (e.g. Ashcraft, 2002; Dew et al., 1984; Hembree, 1990). Furthermore, scores on mathematics anxiety instruments are better predictors than scores on test anxiety instruments of performance on mathematics tests and courses, and the negative thoughts which occur during mathematics problem-solving (Ashcraft & Faust, 1994), indicating that mathematics anxiety is not a component of test anxiety, but that test anxiety likely is a component of mathematics anxiety. This implies that reducing test anxiety may help to reduce mathematics anxiety somewhat, especially when being evaluated in mathematics, but will likely not impact the anxiety around learning and doing mathematics. Alternatively, un-timed assessment strategies should be considered to assuage test anxiety.

Mathematics Anxiety and Pre-service Teachers

Many studies have reported high incidences of mathematics anxiety among primary pre-service teachers (Harper & Daane, 1998; Hembree, 1990; Kelly & Tomhave, 1985; Novak & Tassell, 2017; Sloan, 2010). In a meta-analysis of 151 studies, Hembree (1990) found that pre-service teachers maintain the highest levels of mathematics anxiety when compared to other undergraduate university students. The mathematics anxiety of future teachers is of concern as they are likely to carry this anxiety into their careers, and several studies have suggested that schools, and teachers in particular, play a significant role as environmental antecedents for mathematics anxiety (Beilock & Maloney, 2015; Buckley et al., 2016; Harper & Daane, 1998; Sloan, 2010; Stoehr, 2017; Whyte & Anthony, 2012), and that mathematics anxiety can be passed on from teachers to students (Bulmahn & Young, 1982; Bush, 1989; Çatlioğlu, Birgin, Coştu, & Gürbüz, 2009; Gresham, 2018; Herts & Beilock, 2017; Vinson, 2001; Wood, 1988).

It is a matter of judgement as to whether the level of mathematics anxiety in the population of primary teachers is higher than that in the general population. However, if primary teachers were rated as approximately the same as the general public, but these levels indicate an overall anxiety or dislike for mathematics, then there is still a problem. This is because, as Wood (1988) explains, "elementary teachers are charged with an extremely important role and that role is to engender an excitement for learning in all subject areas, including mathematics" (p. 11).

Hadley and Dorward (2011) investigated mathematics anxiety as a determining factor in the year level primary teachers choose to work in and found that teachers with lower anxiety about mathematics were more commonly teaching in the upper primary level. This corresponds with the author's experience when eliciting pre-service teachers' choices (and reasons for those choices) for lower grade levels for practicum placements.

Recently, pre-service teachers at the two generalist primary teaching programmes in the UAE, self-reported moderate anxiety regarding mathematics, using the Anxiety for Mathematics Scale (McMinn, 2019, see Measuring Anxiety section of this chapter). These results indicated a lower level of anxiety than anticipated by the author, based on experience and past research. However, as suggested previously, it is noted that any mathematics anxiety in pre-service teachers is cause for concern as it has been negatively related to teacher behaviour and student achievement (Haciomeroglu, 2014; Hadley & Dorward, 2011; Hembree, 1990; Muijs & Reynolds, 2015; Peker & Ulu, 2018) and can perpetuate a negative mathematics anxiety cycle.

Measuring Anxiety

For as long as mathematics anxiety has been a topic of interest, researchers have been trying to assess and measure it. Several instruments have been developed, largely using questionnaires and ratings scales. Dreger and Aitkin developed three items for a Number Anxiety questionnaire, which was used in conjunction with an instrument that tests anxiety as a personality trait (Taylor Manifest Anxiety Scale, J. A. Taylor, 1953) and other measures. Two of these items referred to nervousness or 'freezing up' when doing mathematics, and one related to how good participants thought they were at mathematics in relation to other subjects, which may relate more to self-efficacy than mathematics anxiety.[1]

The Mathematics Anxiety Rating Scale (Richardson & Suinn, 1972) has been extensively used (see, for example, Bessant, 1995; Kelly & Tomhave, 1985; Plake & Parker, 1982; Rounds & Hendel, 1980), and has led to the development of the Abbreviated Mathematics Anxiety Rating Scale (Alexander & Martray, 1989), the Revised Mathematics Anxiety Survey (Bursal & Paznokas, 2006) and the 30-item Mathematics Anxiety Rating Scale (Suinn & Winston, 2003), as well as other shortened versions (Plake & Parker, 1982; Rounds & Hendel, 1980). The Fennema–Sherman Mathematics Attitude Scales, developed in 1976, have also been widely used. Only one of the nine scales developed relates to mathematics anxiety, often referred to as the Mathematics Anxiety Scale (MAS), and this scale is often used exclusively, or in conjunction with some of the other attitude scales (Table 5.1).

These instruments have been successfully tried and tested for validity, reliability and internal consistency by the respective authors, and others since. However, as questionnaire instruments collect self-reported data, which cannot be independently verified, the data therefore must be considered at face-value, which could be considered a limitation. Some studies have attempted to overcome this issue by using physiological or neurological measures of anxiety when participants are exposed

[1] Mathematics anxiety and mathematics self-efficacy have repeatedly been found to have a strong, negative correlation, however, are considered distinct constructs.

Table 5.1 *Overview of some mathematics anxiety instruments*

Instrument	Scales	Description
Mathematics anxiety rating scale (MARS) (Richardson & Suinn, 1972)	• Mathematics test anxiety • Numerical anxiety	The 98 items are responded to using a 5-point Likert scale (not at all anxious—very much anxious) *Example items include*: • Adding two three-digit numbers while someone looks over your shoulder • Thinking about an upcoming math test 1 day before
Fennema–Sherman mathematics attitude scales (FSMAS) (Fennema & Sherman, 1976)	• Attitude towards the success in mathematics • Mathematics as a male domain, • Mother (perception of mother's attitudes toward one as a learner of mathematics) • Father (perception of father's attitudes towards one as a learner of mathematics) • Teacher (perception of teacher's attitudes towards one as a learner of mathematics) • Confidence in learning mathematics • *Mathematics anxiety* • Effectance motivation in mathematics • Mathematics usefulness	Each scale has 12 items responded tousing a 5-point Likert scale (from strongly disagree to strongly agree) *Example items from the mathematics anxiety scale include*: • My mind goes blank and I am unable to think clearly when doing mathematics • Math doesn't scare me at all

(continued)

Table 5.1 (continued)

Instrument	Scales	Description
The abbreviated mathematics anxiety rating scale (A-MARS) (Alexander & Martray, 1989)	• Math test anxiety • Numerical anxiety • Math course anxiety	The 25 items are responded to using a 5-point Likert scale (not at all anxious - very much anxious), *Example items include:* • Taking math section of the college entrance exam • Realising you have to take a certain number of math classes to fulfil requirements • Reading a cash register receipt after your purchase
The 30-item MARS scale (Suinn & Winston, 2003)	• Mathematics test anxiety • Numerical anxiety	The 30 items are responded to using a 5-point Likert scale (not at all anxious—very much anxious) *Example items include*: • Getting ready to study for a mathematics test • Dividing a five-digit number by a two-digit number in private with pencil and paper
Revised mathematics anxiety survey (R-MANX) (Bursal & Paznokas, 2006)	N/A	The 30 items are responded to using a 5-point Likert scale (never—always) *Example items include:* • When I hold a math textbook to study I start feeling stomach ache • I can reject helping a child with his homework because I am afraid of facing a question which I cannot solve

(continued)

Table 5.1 (continued)

Instrument	Scales	Description
Anxiety for Mathematics Survey (AMS) (McMinn, 2019)	• Anxiety caused by mathematics learning • Anxiety caused by mathematics evaluation • Anxiety caused by numerical tasks * Anxiety caused by mathematics in real-life situations • Anxiety caused by non-mathematics situations	The 40 items are responded to using a five-point frequency scale (*'how anxious you have felt in the last year'*; not at all anxious, a little anxious, somewhat anxious, anxious and very anxious) *Example items include*: • Being asked to solve word problems in mathematics class • I take an examination (quiz) in a mathematics course • I add up 976+777 on paper • Working out how much something will cost me when there is a '25% off' sale • Being given a "pop" quiz in an Arabic class

to mathematical stimuli. Such studies have assessed mathematics anxiety through measuring heartrates (Dew et al., 1984), stress or emotional responses through sweat gland activity and cortisol secretion (Dew et al., 1984; Dreger & Aitkin, 1957; Pletzer et al., 2010; Mattarella-Micke et al., 2011), and brain activity, through Event-Related Potentials recordings (Núñez-Peña & Suárez-Pellicioni, 2015); and functional Magnetic Resonance Imaging (Lyons & Beilock, 2012; Pletzer et al., 2015; Young et al., 2012). Such measures are likely to identify responses beyond what may be consciously known and, although more costly and complex, could provide a means to triangulate self-reported data.

More recently, the Anxiety for Maths Survey (AMS) was developed specifically for use with pre-service participants in the UAE, providing a contemporary tool fit for use with English as additional language learners. This instrument drew inspiration from existing questionnaires (MARS, A-MARS and R-MANX), but items were modified or developed for language clarity and context. This tool consisted of 40 items across five scales (see Appendix A for sample items) and included a five-point frequency-response format which allowed participants to rate their anxiety (*'how anxious you have felt in the last year'*); not at all anxious, a little anxious, somewhat anxious, anxious and very anxious. The results of the factor analysis, internal consistency and discriminant validity supported the reliability and validity of the data collected from the AMS when used in this context.

Causes of Mathematics Anxiety

Many causes of mathematics anxiety have been suggested over the last six decades, with antecedents categorised into three broad forces (dispositional, environmental and situational), and past educational experiences taking a sizeable amount of blame. As the participants of the current study are future teachers and managers of learning environments, it is important to understand the potentially cyclical effects of how the learning environment affects, and is affected by, mathematics anxiety to reinforce the importance of identifying potential teachers with anxiety with the intention to address the issue, so that it is not perpetuated. As gender is one of the factors that has received most studies regarding mathematics anxiety, this will also be discussed.

Antecedents of Mathematics Anxiety

Researchers have identified three overarching forces considered to be antecedents of mathematics anxiety: dispositional, environmental and situational forces. This section defines each of these.

Dispositional forces, such as personality, are internal to the individual, and include mathematics confidence (Byrd, 1982), self-esteem (Gresham, 2008; Tobias, 1980), self-blame (Byrd, 1982), concern over the perceived disapproval of significant others (Byrd, 1982; Tobias, 1980), lack of confidence or persistence (Fennema &

Sherman, 1976; Hadfield & McNeil, 1994), attitude towards mathematics (Byrd, 1982), tendency towards perfectionism (Walsh & Ugumba-Agwunobi, 2002) and learning styles (Hodges, 1983).

Environmental forces are external to the individual and may include: social-economic status (Byrd, 1982); the influence of parental perceptions, particularly those of mothers (Buckley et al., 2016; Byrd, 1982; Sloan, 2010; Whyte & Anthony, 2012); the influence of teachers (Buckley et al., 2016; Whyte & Anthony, 2012); classroom environment (Dossel, 2016; Trujillo & Hadfield, 1999) and the wider social context (Trujillo & Hadfield, 1999). Both environmental and dispositional forces have been linked to trait anxiety (Buckley et al., 2016).

Alternatively, situational forces have been linked to state anxiety (Buckley et al., 2016). Situational forces are the contextual factors that underlie the experience of mathematics anxiety and are distinct from environmental forces with respect to the time frame, with environmental factors occurring in the past and situational factors involving those that are current. Situational antecedents may include instructional methods, including teacher-related factors (Byrd, 1982; Tobias, 1980), mathematics being presented as rigid sets of rules (Hadfield & McNeil, 1994; Kogelman, 1978; Lazarus, 1974), the conventions of mathematical notation, ambiguous terminology and emphasis on a single correct answer (Byrd, 1982; Tobias, 1980) and rote learning (Bulmahn & Young, 1982). Situational forces for UAE pre-service teachers would include the mathematics classroom environments at the higher educational institutes in which they are studying, and other learning environments, such as practicum experiences.

Teachers and Teaching

That the primary mathematics classroom may be a beginning point for mathematics anxiety has been proposed several times over the last 60 years (Beilock & Maloney, 2015; Brady & Bowd, 2005; Buckley et al., 2016; Gough, 1954; Hadfield & Lillibridge, 1991; Harper & Daane, 1998; Hilton, 1980; Sloan, 2010; Stoehr, 2017; Trujillo & Hadfield, 1999; Whyte & Anthony, 2012; Wood, 1988). As pre-service teachers may bring mathematics anxiety based on their own past school experiences with them, and because mathematics anxious teachers tend to deploy ineffective teaching methods and may pass on their anxiety (see Consequences of Mathematics Anxiety section), it is important to understand the influence of teachers on mathematics anxiety to further highlight the importance of identifying mathematics anxious teachers. This section will discuss the impact of the learning environment as it relates to mathematics anxiety, the role the teacher plays within it and teaching methods that may cause and exacerbate anxiety.

The learning environment plays a significant role as an environmental antecedent for mathematics anxiety. In the early 1980s, Fraser and Fisher (1982) highlighted that student perceptions of classroom environments were an important variable responsible for differences in a variety of student outcomes. An important characteristic of any learning environment is the subjective feelings, emotions and attitudes students bring with them into the classroom concerning a particular subject area. As Taylor and Fraser (2013, pp. 299–300) assert, "While feelings of joy and enjoyment are certainly helpful and welcomed in the classroom, feelings of fear and dread seem to be a part of some classrooms where some subjects, especially mathematics, are taught". Indeed, student perceptions of elements of the learning environment have been associated with anxiety in learning. In particular, perceived punitive teacher behaviour (Frenzel et al., 2007; Helmke, 1983), achievement pressure (Goetz et al., 2006; Helmke, 1983) and competition within the class (Frenzel et al., 2007) have been positively related to anxiety. However, when students perceived the quality of instruction to be high, anxiety was slightly reduced (Frenzel et al., 2007). Research specifically looking at mathematics anxiety in 745 secondary students and the learning environment, found that students were less anxious about the mathematics classroom and mathematics learning when there was more peer interaction and acceptance, and more motivation and time on task (Taylor & Fraser, 2013).

Several learning environment studies have identified the teacher as the most crucial element in a mathematics classroom (Aldridge et al., 1999; Bekdemir, 2010; Tobias, 1980). Byrd (1982) found that teacher personality and an uncomfortable classroom atmosphere contributed to making a mathematics learning situation anxiety-provoking. As Byrd (1982, pp. 176–177) stated, "... an individual with a lot of self-doubt and a high need for approval may experience anxiety in the presence of an aloof teacher because approval needs are not met, and self-esteem is threatened".

Uusimaki and Nason (2004) reported that their participants attributed most of their mathematics anxiety to prior school experiences, with 72% of the reasons particularly ascribed to primary school teachers. These past negative experiences seem so insidious that Kogelman and Warren defined mathematics anxiety as, "an intense emotional reaction to math based on *past experiences*" (1978, p. 10; emphasis added); and Freedman defined mathematics anxiety as "an emotional reaction to mathematics based on a *past unpleasant experience* which harms future learning" (2003, p. 1; emphasis added).

Often this blame is linked to teachers themselves (Buckley et al., 2016; Byrd, 1982; Gough, 1954; Tobias, 1980; Uusimaki & Nason, 2004; Wood, 1988), with study participants citing embarrassment in front of others when struggling with problems and being kept in after school (Byrd, 1982), doing mathematics at the blackboard alone (Tobias, 1980), verbal, or even physical abuse in mathematics classrooms (Afflack, 1978; Levine, 1996; Wood, 1988). This could be because the teachers themselves experienced mathematics anxiety, which can lead to unfriendly or defensive responses to students' own difficulties with the subject. Some adults have indeed reported that they developed mathematics anxiety in direct response to a teacher in their past reacting angrily when they asked for help or seemed insensitive towards their struggle with maths (Jackson & Leffingwell, 1999).

Traditional teaching styles, such as is still prevalent in many parts of the UAE (Kippels & Ridge, 2019; Langton, 2017), are also highlighted as potential causes of mathematics anxiety, including the dependence on rote learning and memorisation, the emphasis on getting the right answer and using the right method, unrealistic applications or problems, ability grouping and the impression that mathematics is rigid and authoritative (Byrd, 1982; Harper & Daane, 1998; Hilton, 1980; Kelly & Tomhave, 1985; Stoehr, 2017; Tobias, 1980). This seems to be supported by research that found when non-traditional methods are used, anxiety can be reduced (Tobias, 1980). Teaching practices such as the assumption that mathematical processes and procedures were simple and self-explanatory, and the use of mathematics vocabulary without appropriate clarification, may also lead to the belief that mathematicians are born with a gift for it, leaving mathematics anxious individuals struggling, frustrated and believing they simply do not have a mathematics mind (Byrd, 1982; Cornell, 1999; Gough, 1954; Kogelman & Warren, 1978).

It is also believed that the successive nature of mathematics may contribute to mathematics anxiety (Byrd, 1982; Gough, 1954; Lazarus, 1974). As many mathematics skills are accumulative and sequential, any class time missed could have detrimental effects. There are numerous reasons students may be absent from school for days at a time, and distractions at home and inadequate teaching may be additional reasons why a topic may be missed. This suggests it is almost inevitable that young students will miss an important stage upon which at least part of the future development of the subject depends. "Whatever the reason, mathematics anxious individuals often seem to relate their anxiety to this feeling of loss due to misunderstanding of a prerequisite topic" (Byrd, 1982, p. 56).

Mathematics Anxiety and Gender

Repeatedly, studies[2] have shown no difference between male and female students in mathematics performance (see, for example, Harari, 2013; Reilly et al., 2019), yet female students report higher mathematics anxiety (see, for example, Devine et al., 2012; Else-Quest et al., 2010; Hembree, 1990; Wigfield & Meece, 1988). These gender differences seem to develop in adolescence and may be due to gender stereotypes of mathematics traditionally being a male dominant field or may have been 'caught' from mathematics anxious female teachers as discussed above. However, some studies have shown that females, in general, exhibit higher levels of trait anxiety than males (see for example, Bieg et al., 2015; Chapman et al., 2007; Costa et al., 2001; Goetz et al., 2013) and have been found to show greater anxiety than males even in subjects where their actual performance is often higher than that of males, such as foreign language learning (Park & French, 2013). This is supported by the self-reported anxiety of pre-service teachers in the UAE, the vast majority (95%) of whom are female, who indicated similar anxiety for other subjects and activities required for their degree on the scale with analogous situations (AMS).

Devine et al. (2012) found that mathematics anxiety affected mathematics performance in females only, even after controlling for general test anxiety. In their study

[2] In countries where education is provided equally for both genders.

with 11–15-year-olds, they found that male and female students performed comparably in mental mathematics tests, but girls had both higher mathematics anxiety and higher test anxiety. Positive correlations between mathematics anxiety and test anxiety and negative correlations between mathematics anxiety and mathematics performance were found for both male and female students, and both male and female students showed a negative correlation between mathematics anxiety and mathematics performance. However, regression analyses showed that for male students, this relationship disappeared after controlling for general test anxiety. Only girls continued to show an independent relationship between mathematics anxiety and mathematics performance. These results suggest that females may have had the potential to perform better than males in mathematics, but their performance may have been tempered by their higher levels of mathematics anxiety.

When considering the mathematics anxiety of pre-service teachers, issues pertaining to gender are of importance considering the relative numbers of men and women enrolling in teacher education programmes. There have been a range of factors posited for the gender imbalance in the teaching profession, not least the societal perception that primary teaching is 'women's work'. A downward trend in the percentage of primary teachers who are male has been observed in numerous countries such as England, the United States of America, Australia and New Zealand (Australian Bureau of Statistics, 2019; Bureau of Labor Statistics, 2013; Cushman, 2008; de Salis et al., 2019). In the UAE, the Abu Dhabi Educational Council (ADEC) announced plans in 2011 to feminise primary schools in the emirate (Dickson & Le Roux, 2013). Since that time, the only teacher education programme that admitted male students was closed.

While many studies of mathematics anxiety involving pre-service teachers do not compare results on the basis of gender, a quick look at the gender makeup of the participants reflects the gender imbalance. Given that females have been repeatedly found to have higher levels of mathematics anxiety and that mathematics anxious female teachers can pass on this anxiety to their female students, mathematics anxiety is very much a gender-related issue.

Consequences of Mathematics Anxiety

Gresham (2008) claimed that, at best, mathematics anxiety can affect the learning capabilities of students in the short term and, at worst, can develop into the more serious mathematics avoidance thereby limiting options in the long term. A negative correlation between mathematics anxiety and mathematics achievement has been repeatedly found in studies with all levels of students from primary to college (see, for example, Ashcraft & Kirk, 2001; Betz, 1978; Devine et al., 2012; Hembree, 1990; Ma & Xu, 2004; Pourmoslemi et al., 2013; Ramirez et al., 2013; Tocci & Engelhard, 1991; Wigfield & Meece, 1988). If pre-service teachers bring with them anxiety for mathematics, it could affect the learning of, or performance in, mathematics during teacher education, which could, in turn, affect teacher efficacy, the belief in one's competence to successfully accomplish teaching tasks. Teachers with mathematics anxiety are likely to use more traditional, surface-level teaching and can even perpetuate mathematics anxiety in their students. Therefore, it is important to identify mathematics anxiety in pre-service teachers, so that it may be addressed. This section describes the typical teaching methods of mathematics anxious teachers, and the transferable nature of mathematics anxiety.

Avoidance

Although there is insufficient evidence that a causal link between mathematics anxiety and mathematics avoidance exists (Kelly & Tomhave, 1985), many authors have asserted that students with mathematics anxiety are less likely to persevere with mathematics learning, demonstrate low independence, and a desire to avoid mathematics (see, for example, Bessant, 1995; Buckley et al, 2016; Chinn, 2012; Gresham, 2008; Hembree, 1990). Mathematics anxiety has been said to affect the amount of effort one is willing to exert to learn mathematics (Chinn, 2012; Fennema & Sherman, 1976; Muis, 2004), and sufferers are likely to develop negative attitudes towards the subject (Ashcraft & Moore, 2009; Hembree, 1990; Ma, 1999). It is possible that this has a snowball effect. Morris (1981) proposes that the fatalistic attitude of mathematics anxious students can become a self-fulfilling prophecy. And indeed, such anxiety seems to have a large adverse influence on the desire to enrol in mathematics courses beyond minimum requirements in secondary school, particularly for female students (Devine et al, 2012; Fennema & Sherman, 1976). College students with mathematics anxiety are more inclined to drop out of mathematics courses (Ma & Xu, 2004). Tobias (1980) believes that students who avoid mathematics in high school and college are typically manifesting a relatively high degree of mathematics anxiety at the same time. Some mathematics anxious students may avoid majors or even entire colleges with heavy mathematics requirements (Beilock & Maloney, 2015; Byrd, 1982; Hembree, 1990; Ma, 1999; Mack & Walsh, 2013; Richardson & Suinn, 1972). This is clearly a concern when science, technology, engineering and mathematics (STEM) subjects are being given priority in UAE education, as the country moves towards a knowledge-based economy and away from reliance on the oil economy (Arab Knowledge Report, 2011). Beyond education, mathematics anxiety is reported to have an impact on the choices for careers, where the sciences and other careers that use quantitative skills, and even jobs working with computers are avoided (Byrd, 1982; Devine et al, 2012; Hembree, 1990; Ma, 1999). This is naturally of concern in our increasingly technological society. Over thirty years ago, Tobias declared that it is a myth that anyone can avoid mathematics and still live a productive intellectual and professional life:

> Mathematics is no longer just an entry-level prerequisite for engineering, the physical sciences, and statistics. Its principles and techniques, along with computers, have become part of almost all areas of work, and its logic is used in thinking about almost everything. (Tobias, 1990, p. 47)

Traditional Teaching Methods

Traditional teaching methods have already been mentioned as a cause of mathematics anxiety, but also appear to be a consequence of mathematics anxiety, thus maintaining the negative cycle. When mathematics anxious individuals become teachers, they can bring with them a dislike of teaching mathematics (Gresham, 2018; Ma & Xu, 2004). Teachers with mathematics anxiety are likely to teach mathematics at a surface level, using more traditional teaching methods (that reform efforts in the UAE are trying to move away from), such as whole class, lecture-style lessons, rote memorisations of algorithms, and fewer problem-solving techniques and games, possibly disregarding students' preferred learning styles (Bekdemir, 2010; Bursal & Paznokas, 2006; Bush, 1989; Gresham, 2018; Hadfield & McNeil, 1994; Trujillo & Hadfield, 1999). It has also been found that mathematics anxious teachers avoid mathematics, spending less time on planning for mathematics and dedicating less teaching time to the subject (Bush, 1989; Levine, 1993; Peker & Ertekin, 2011; Swetman et al., 1993); in some cases, 50% less time teaching mathematics than non-anxious teachers (Schmidt & Buchmann, 1983). This, naturally, causes reservations about the effectiveness of such teachers. Academic success in mathematics classes of teachers with mathematics anxiety is therefore likely to suffer, and as the above teaching methods have also been highlighted as a cause for mathematics anxiety, it is possible that a negative cycle is being perpetuated.

Passing It on

It has been suggested that teachers pass on their avoidance of the subject (Bulmahn & Young, 1982; Kelly & Tomhave, 1985; Lazarus, 1974), and that, in fact, mathematics anxiety can be passed from the teacher onto the students (Bulmahn & Young, 1982; Bush, 1989; Çatlioğlu et al., 2009; Gresham, 2018; Vinson, 2001; Wood, 1988), potentially creating another generation of mathematics anxious people.

When the mathematics anxiety of 17 female first and second grade teachers was assessed, and correlated with the achievement of their students at the beginning and end of the school year, it was found that the female teachers with high mathematics anxiety had a large influence on girls' gender-related beliefs about who is good at mathematics, which in turn negatively affected the girls' mathematics achievement (Beilock et al., 2010). The implication here is that if these teachers were just ineffective at teaching mathematics, a relationship between teacher anxiety and the mathematics achievement of both boys and girls could be expected, but this was not the case. This is a disturbing finding at such an early stage in girls' educational careers, especially because Ma and Xu (2004) found that for girls, mathematics anxiety has the tendency to endure over time once formed. Similarly, Suman et al. (2015) in their study with 138 pre-service primary school teachers found that their mathematical fear resulted in no small part from the attitudes and behaviours of their teachers.

In contrast, some mathematics anxious primary teachers have expressed an eagerness to break the cycle and to reduce mathematics anxiety in their own students.

Such pre-service teachers were explicitly using their negative prior experiences with mathematics during their own schooling as the motivation for teaching mathematics better to their future students (Brown et al., 2012; Widmer and Chavez (1982) . This is encouraging, but to date, no study has followed up to see whether pre-service teachers were able to overcome their anxiety to an extent that their attitudes and teaching approaches did not create or increase mathematics anxiety in their students. These findings do, however, highlight the importance of identifying the mathematics anxiety of pre-service teachers, with an aim to channel the anxiety into more positive outcomes.

Breaking the Cycle

UAE national pre-service teachers from two higher education institutes offering generalist primary teaching programmes (described above) were invited to complete the AMS questionnaire. The two institutes provided a total target sample of approximately 550 UAE national pre-service teachers, predominantly female; 176 students participated. This section presents the results from the AMS questionnaire and discusses possible interventions and other recommendations for breaking the mathematics anxiety cycle.

Results of Anxiety for Mathematics Survey

The means for all five scales fell between 2.05 and 2.95 (using a 5-point Likert scale), and below the midpoint of 3.00 (McMinn, 2019), see box and whisker plot below. Based on the descriptors used in the response format, participants rated their mathematics mean anxiety to be a 'little' to 'somewhat' anxious. Pre-service teachers rated their Anxiety caused by Mathematics Evaluation the highest (mean = 2.95) for the five scales. For all five scales, some participants were 'anxious' or 'very anxious' about mathematics, with at least 50% of participants indicating that they were a least a little anxious for all five scales (Fig. 5.1).

Interventions

Given that mathematics anxiety has been related to mathematics teaching anxiety, teaching efficacy, beliefs about mathematics, teacher behaviour and student achievement, the finding that UAE national pre-service teachers have moderate mathematics anxiety may have negative implications for the future of mathematics education in UAE primary schools, unless addressed. Fortunately, some past studies have shown positive results from interventions and others have offered suggestions for reducing mathematics anxiety, either in general, or specifically for mathematics anxious pre-service teachers. This section discusses these interventions, what strides have been

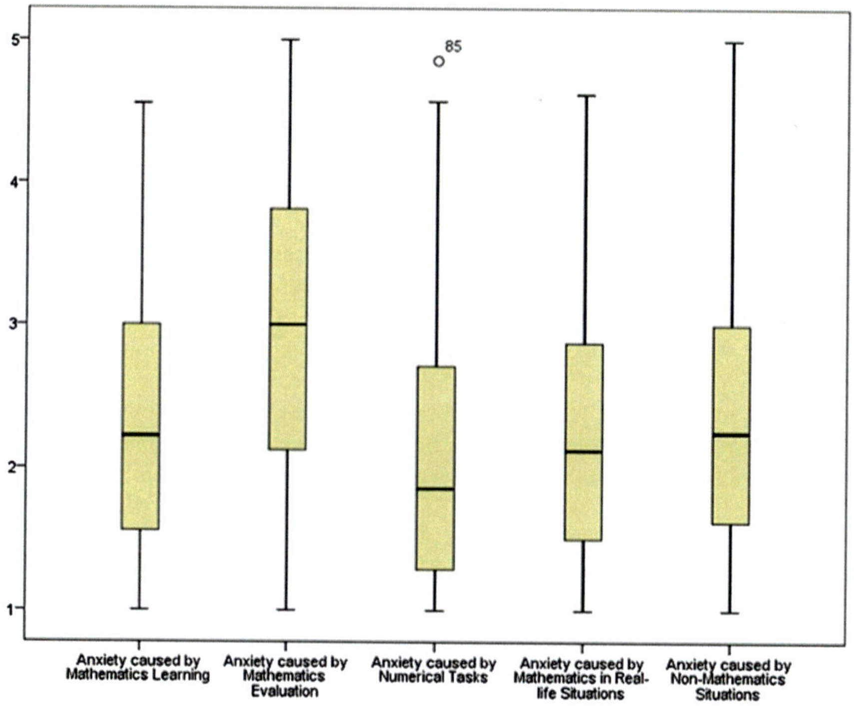

Fig. 5.1 *Box and whisker plot for scales of the Anxiety for Mathematics Survey (AMS)*

taken in a UAE teacher education programme thus far, and recommendations for other teacher education programmes.

Behavioural Methods

In a synthesis of education, psychology and neuroscience research literature, Buckley et al (2016) found evidence that suggested methods such as relaxation, mindfulness, cognitive reappraisal and cognitive behavioural approaches may be effective in overcoming mathematics anxiety. Similarly, Hembree's (1990) meta-analysis of 151 studies also found that the largest reductions in mathematics anxiety were linked to behavioural methods, that aimed to relieve 'emotionality' or feelings of dread and nervous reactions, for example, through systematic desensitisation; and cognitive behavioural treatments which addressed the worry factor, but also offered help in reducing emotionality. Treatments that resulted in significant mathematics anxiety reduction were accompanied by significant increases in mathematics test scores (Hembree, 1990), adding evidence to the argument that mathematics anxiety causes lower achievement.

User-Friendly Curriculum

However, interventions need not be psychological. Educational activities such as adopting 'user-friendly' mathematics curricula, emphasising problem-solving, using hands-on activities and working cooperatively in small groups were found to reduce mathematics anxiety in a study with high school students (Ma & Xu, 2004). They also suggested that, for boys, improving their mathematics achievement reduces their mathematics anxiety, intimating the deficits model; whereas for girls, it is more important to prevent mathematics anxiety forming (as it tends to be long-lasting), which often happens in critical periods such as school transition periods. In a study with college students, Betz (1982) found that the more prior mathematics preparation a student had, the less likely he or she was to report high levels of mathematics anxiety. This indicates that, if mathematics anxiety begets mathematics avoidance, more mathematics courses during high school that utilise the methods listed above, more opportunities for success during those courses, and careful monitoring of girls in mathematics courses around periods of transition, would go a long way to reducing anxiety for college students. Assuming mathematics anxiety is already formed, Tobias suggests the development of mathematics workshops. These should utilise mathematics instructors and counsellors in tandem, allow students to learn new mathematics material in a slow-paced environment in which they are encouraged to reflect on their affective state during mathematics activities, and for the development of study skills from passive to active, can go a long way in offering "mathematical mental health" (1990, p. 49).

Teacher Education Programmes

Early intervention of mathematics anxiety with pre-service teachers is likely to have significant benefits (Buckley et al., 2016), especially as mathematics anxious teachers spend less time planning and teaching mathematics and teach in ways that have been linked with cultivating mathematics anxiety in their students. Therefore, it is recommended that a component of all teacher education programmes be explicitly aimed at addressing mathematics anxiety (as purported by Gresham, 2007; Harper & Daane, 1998), and several strategies for doing so have been suggested. Openly addressing students' attitudes towards mathematics (L. Taylor & Brooks, 1986) and providing opportunities for pre-service teachers to reflect about any anxiety or negative attitudes through writing about mathematics, sharing experiences and discussing the implications (Harper & Daane, 1998), could serve to reduce anxiety. It is recommended that teacher education programmes regularly monitor the levels of pre-service teachers' mathematics anxiety to ensure they are being reduced and offer extra support for those who maintain higher anxiety levels. In the largest teacher education provider in the UAE, an explicit discussion of students' feelings and beliefs about mathematics, good and bad experiences in their past learning and discussions about what kind of teachers they want to be, have been built into a mathematics education course. It is hoped that this will bring anxiety to the fore to be addressed. While descriptive statistics regarding the mathematics anxiety of UAE pre-service teachers have not been collected since these interventions were implemented, anecdotal comments from students suggest they are, at least, a good start.

Wood (1988) investigated several programmes specifically designed to reduce primary pre-service teachers' mathematics anxiety and found that programmes that improved attitudes towards mathematics and lowered mathematics anxiety all had in common one thing: the students were taught well. These programmes used effective teaching techniques, such as, not assuming prior knowledge, teaching new material slowly and encouraging students to talk through their thought processes. Tobias (1990) suggests similar techniques, through the development of mathematics workshops, ideally utilising mathematics instructors and counsellors in tandem, which involves students reflecting on their mathematics learning and developing a willingness to learn. It is recommended that pre-service teachers are taught in mathematics classes by highly competent and confident instructors, which may simply mean careful scheduling and allocating of courses in situations where any faculty could be assigned mathematics courses.

Mathematics methods courses can also act as an intervention, as learning to teach mathematics during pre-service teacher training seems to reduce mathematics anxiety (Buckley, 2016; Gresham, 2007; Hadley & Dorward, 2011; Harper & Daane, 1998; Sloan, 2010; Tooke & Lindstrom, 1998). Furthermore, having extensive fieldwork experiences and familiarising primary pre-service teachers with the world of mathematics teaching (Harper & Daane, 1998; Tooke & Lindstrom, 1998) have also been suggested as a method for lowering mathematics anxiety. Therefore, it is recommended that teacher education programmes increase fieldwork experiences. This could be in the form of additional practicum experiences, micro-teaching and/or observations of model lessons by in-service teachers, and these experiences should occur early in the programme, rather than waiting until the end. In the two teacher education programmes surveyed in the UAE, students complete practicum placements either once a year or every semester. However, this could be further supported with additional micro-teaching and observations.

Sloan (2010) found that the majority of early childhood and primary pre-service teachers she interviewed indicated that the use of manipulatives in their mathematics methods course reduced mathematics anxiety, which may be useful in the current context given the moderate anxiety reported. The participants claimed the manipulatives facilitated conceptual understanding of mathematics, which enhanced their confidence and improved their attitudes towards mathematics. Other studies with early childhood and/or primary pre-service teachers and mathematics anxiety found similar results regarding the use of manipulatives (Gresham, 2007; Harper & Daane, 1998; Vinson, 2001). It is recommended that all teacher education programmes incorporate the use of manipulatives into both content and methodology courses wherever possible to enhance learning and provide a model for best practice. Manipulatives should be used in the students' own learning, and as an overt model for their future teaching. These manipulatives should be comparable to those which the pre-service teachers will find in a standard classroom during teaching practicum, so they are familiar, relatable and useful. In a UAE B.Ed. programme, opportunities for incorporating manipulatives into mathematics classes had been planned prior to the COVID-19 pandemic; however, efforts to implement these activities have been hampered with the move to online learning due to the pandemic.

Relating mathematics experiences to real-life contexts has also been espoused as reducing mathematics anxiety in early childhood and primary pre-service teachers. (Gresham, 2007). It is recommended that mathematics courses in teacher education programmes incorporate links to real-life wherever appropriate. One obvious example is the frequent use of geometric patterns in Islamic art, and with the opening of the Louvre in Abu Dhabi, and the planned Guggenheim Museum, this could be extended into a variety of art forms. Simple budgeting is another relevant way to make connections to real-life, for example, pre-service teachers could investigate costs related to pre- and post-paid mobile phone plans, or which campus coffee shop offers best value for money. Pre-service teachers in a UAE teacher education programme were tasked with developing and conducting a mathematics investigation as part of an assessment activity. The focus of their investigation was of their own choosing (in discussion with their course teacher) and ranged from exploring the best value for money salons to potential small business opportunities on campus.

As the mean for the 'Anxiety caused by Evaluation in Mathematics' scale was the highest reported by UAE pre-service teachers (using the AMS), at 2.95, this is an area that UAE teacher educators should consider addressing. The development of effective study skills could help. It is possible that this could be accomplished through games and activities, and with lots of practice in simulated evaluative situations with the material, and progressing in difficulty. Additionally, a longitudinal study involving pre-service teachers ($n = 20$) in a B.Ed. programme in the United Kingdom found that college mathematics classes helped alleviate anxiety by promoting mathematical processes and reducing the focus on getting the right answer. Ramirez and Beilock (2011) tested an intervention for mathematics anxiety that amounts to 'writing out' the negative, anxious thoughts. Both their laboratory and field experiments showed that writing expressively for 10 min prior to an exam significantly improved performance compared to a control condition (e.g. writing about untested exam material). This could be because the 'writing out' of worries frees working memory resources previously expended on anxiety, which can be then used towards the task at hand. Interestingly, this intervention seemed to only benefit individuals with high test anxiety, possibly because those with lower anxiety have less worries that need to be written out. This has exciting potential for reducing the disadvantage mathematics anxious individuals may face in exam situations. In the largest UAE teacher education provider, exams for mathematics courses have been removed and replaced with assignments. This removes the time pressure element which may exacerbate the evaluation element of mathematics anxiety and allows students time to draft work, get feedback, ask questions, etc. Where teacher education programmes cannot do this, it is suggested that 10 minutes of expressive writing time at the beginning of an exam could be easily facilitated.

The Learning Environment

The learning environment should also be carefully considered. A recent study in the UAE relating the learning environment and mathematics anxiety of pre-service teachers found that the more opportunities pre-service teachers are given to express their opinions, discuss ideas and are listened to by their peers in mathematics classes, the lower their anxiety (McMinn & Aldridge, 2020). Other results from this study, and other past research, also denote a negative correlation between the learning environment and mathematics anxiety, that is, the more favourably students perceive the mathematics learning environment the lower their anxiety. Mathematics learning environments perceived as more favourable by pre-service students in the UAE included those where students felt comfortable, supported and safe, although, interestingly, more collaboration was positively related to higher mathematics evaluation anxiety. This may suggest that students felt more anxious about mathematics evaluation when collaboration was involved, and therefore teacher education programmes may need to plan evaluative tasks thoughtfully.

Conclusion

In the context of ongoing educational reform projects across the UAE, and in view of the aim of The Abu Dhabi Economic Vision Report 2030 to have 90% of employees in the education sector being UAE nationals by 2030 (The Abu Dhabi Government, 2008), and a priority on STEM subjects in the UAE education, it is essential that mathematics anxiety be identified and explicitly addressed. The future primary teachers of the UAE's children have reported moderate (mean) levels of mathematics anxiety, a cause for concern when primary teachers are often cited as the cause of mathematics anxiety.

Relatively simple modifications to teacher education programmes (such as teaching approaches, types of assessments, awareness of anxiety, etc.) could be made to reduce mathematics anxiety. Not only could this have a positive impact on the quality teaching of mathematics in Abu Dhabi schools, and promote the reform efforts, but it is also likely to positively affect teacher retention and contribute to the nation's goals.

Appendix A

Description and Sample Item for Each Anxiety for Mathematics Survey (AMS) Scale

Scale	Purpose	Sample item
Anxiety caused by mathematics learning	To assess the extent to which anxiety is caused by learning mathematics	I feel anxious when walking into a mathematics class
Anxiety caused by mathematics evaluation	To assess the extent to which anxiety is caused by being evaluated in mathematics	I feel anxious when I think about a mathematics test that I have in one hour
Anxiety caused by numerical tasks	To assess the extent to which anxiety is caused by performing numerical tasks	I feel anxious when someone watches me add up a list of numbers
Anxiety caused by mathematics in real-life situations	To assess the extent to which anxiety is caused by undertaking mathematics in real-life situations	I feel anxious when working out how much something will cost when there is a '25% off' sale
Anxiety caused by non-mathematics situations	To assess the extent to which anxiety is caused by analogous situations	I feel anxious when thinking about an upcoming Arabic test 5 min before

References

Afflack, R. (1978). Mathophobia: A mini-course for the mathophobic In J. Jacobs (Ed.), *Perspectives on women and mathematics* (pp. 73–100). Eric Clearinghouse for Science, Mathematics and Environmental Education.

Aldridge, J. M., Fraser, B. J., & Huang, T.-C.I. (1999). Investigating classroom environments in Taiwan and Australia with multiple research methods. *The Journal of Educational Research, 93*(1), 48–62. https://doi.org/10.1080/00220679909597628

Alexander, L., & Martray, C. R. (1989). The development of an abbreviated version of the mathematics anxiety rating scale. *Measurement and Evaluation in Counseling and Development, 22*(3), 143–150.

Arab Knowledge Report. (2011). Preparing future generations for the knowledge society. Mohammed bin Rashid Al Maktoum Foundation. http://www.undp.org/content/dam/rbas/report/AKR2010-2011-Eng-Full-Report.pdf.

Ashcraft, M. H. (2002). Math anxiety: Personal, educational, and cognitive consequences. *Current Directions in Psychological Science, 11*(5), 181–185. https://doi.org/10.1111/1467-8721.00196

Ashcraft, M. H., & Faust, M. W. (1994). Mathematics anxiety and mental arithmetic performance: An exploratory investigation. *Cognition & Emotion, 8*(2), 97–125. https://doi.org/10.1080/02699939408408931

Ashcraft, M. H., & Kirk, E. P. (2001). The relationships among working memory, math anxiety, and performance. *Journal of Experimental Psychology: General, 130*(2), 224–237. https://doi.org/10.1037/0096-3445.130.2.224

Ashcraft, M. H., & Moore, A. M. (2009). Mathematics anxiety and the affective drop in performance. *Journal of Psychoeducational Assessment, 27*(3), 197–205. https://doi.org/10.1177/0734282908330580

Atkinson, R. (1988). *An exploration of the factors relating to the system of mathematics anxiety.* (Doctoral Dissertation). https://search-proquest-com.dbgw.lis.curtin.edu.au/docview/303640816?accountid=10382. ProQuest Dissertations Publishing database. (303640816).

Australian Bureau of Statistics. (2019). *Schools Australia series.* Australian Bureau of Statistics. https://www.abs.gov.au/ausstats/abs@.nsf/mf/4221.0.

Baloğlu, M. (1999). *A comparison of mathematics anxiety and statistics anxiety in relation to general anxiety.* https://files.eric.ed.gov/fulltext/ED436703.pdf.

Bureau of Labor Statistics. (2013). 2013 BLS current population study. http://www.menteach.org/resources/data_about_men_teachers.

Beilock, S. L., & Maloney, E. A. (2015). Math anxiety: A factor in math achievement not to be ignored. *Policy Insights from the Behavioral and Brain Sciences, 2*(1), 4–12. https://doi.org/10.1177/2372732215601438

Bekdemir, M. (2010). The pre-service teachers' mathematics anxiety related to depth of negative experiences in mathematics classroom while they were students. *Educational Studies in Mathematics, 75*(3), 311–328. https://doi.org/10.1007/s10649-010-9260-7

Bessant, K. C. (1995). Factors associated with types of mathematics anxiety in college students. *Journal for Research in Mathematics Education, 26*(4), 327–345. https://doi.org/10.2307/749478

Betz, N. E. (1978). Prevalence, distribution, and correlates of math anxiety in college students. *Journal of Counseling Psychology, 25*(5), 441–448. https://doi.org/10.1037/0022-0167.25.5.441

Bieg, M., Goetz, T., Wolter, I., & Hall, N. C. (2015). Gender stereotype endorsement differentially predicts girls' and boys' trait-state discrepancy in math anxiety. *Frontiers in Psychology, 6*, 1404. https://doi.org/10.3389/fpsyg.2015.01404

Birenbaum, M. (2007). Assessment and instruction preferences and their relationship with test anxiety and learning strategies. *Higher Education, 53*, 749–768. https://doi.org/10.1007/s10734-005-4843-4

Brady, P., & Bowd, A. (2005). Mathematics anxiety, prior experience and confidence to teach mathematics among pre-service education students. *Teachers and Teaching, 11*(1), 37–46. https://doi.org/10.1080/1354060042000337084

Brown, A. B., Westenskow, A., & Moyer-Packenham, P. S. (2012). Teaching anxieties revealed: Pre-service elementary teache's' reflections on their mathematics teaching experiences. *Teaching Education, 23*(4), 365–385. https://doi.org/10.1080/10476210.2012.727794

Brush, L. R. (1981). Some thoughts for teachers on mathematics anxiety. *Arithmetic Teacher, 29*, 37–39.

Buckley, S., Reid, K., Goos, M., Lipp, O. V., & Thomson, S. (2016). Understanding and addressing mathematics anxiety using perspectives from education, psychology and neuroscience. *Australian Journal of Education, 60*(2), 157–170. https://doi.org/10.1177/0004944116653000

Bulmahn, B. J., & Young, D. M. (1982). On the transmission of mathematics anxiety. *The Arithmetic Teacher, 30*(3), 55–56.

Bursal, M., & Paznokas, L. (2006). Mathematics anxiety and preservice elementary teache's' confidence to teach mathematics and science. *School Science and Mathematics, 106*(4), 173–180. https://doi.org/10.1111/j.1949-8594.2006.tb18073.x

Bush, W. S. (1989). Mathematics anxiety in upper elementary school teachers. *School Science and Mathematics, 89*(6), 499–509.

Byrd, P. (1982). *A descriptive study of mathematics anxiety: Its nature and antecedents.* ProQuest Dissertations Publishing.

Çatlioğlu, H., Birgin, O., Coştu, S., & Gürbüz, R. (2009). The level of mathematics anxiety among pre-service elementary school teachers. *Procedi—Social and Behavioral Sciences, 1*(1), 1578–1581. https://doi.org/10.1016/j.sbspro.2009.01.277.

Chang, H., & Beilock, S. L. (2016). The math anxiety-math performance link and its relation to individual and environmental factors: A review of current behavioral and psychophysiological research. *Current Opinion in Behavioral Sciences, 10*(C), 33–38. https://doi.org/10.1016/j.cobeha.2016.04.011.

Chapman, B. P., Duberstein, P. R., Sörensen, S., & Lyness, J. M. (2007). Gender differences in five factor model personality traits in an elderly cohort: Extension of robust and surprising findings to an older generation. *Personality and Individual Differences, 43*, 1594–1603. https://doi.org/10.1016/j.paid.2007.04.028

Chinn, S. (2012). Beliefs, anxiety, and avoiding failure in mathematics. *Child Development Research, 1–8*,. https://doi.org/10.1155/2012/396071

Cornell, C. (1999). I hate math!: I couldn't learn it, and I can't teach it! *Childhood Education, 75*(4), 225. https://doi.org/10.1080/00094056.1999.10522022

Costa, P. T., Terracciano, A., & McCrae, R. R. (2001). Gender differences in personality traits across cultures: Robust and surprising findings. *Journal of Personality and Social Psychology, 81*, 322–331. https://doi.org/10.1037/0022-3514.81.2.322

Covington, M. V., & Omelich, C. L. (1987). "'I knew it cold before the exam'": A test of the anxiety-blockage hypothesis. *Journal of Educational Psychology, 79*, 393–400.

Culler, R. E., & Holahan, C. J. (1980). Test anxiety and academic performance: The effects of study-related behaviors. *Journal of Educational Psychology, 72*, 16–20.

Cushman, P. (2008). So what exactly do you want? What principals mean when they say "male role model." *Gender and Education, 20*(2), 123–136. https://doi.org/10.1080/09540250701805847

Devine, A., Fawcett, K., Szűcs, D., & Dowker, A. (2012). Gender differences in mathematics anxiety and the relation to mathematics performance while controlling for test anxiety. *Behavioral and Brain Functions : BBF, 8*(1), 33–33. https://doi.org/10.1186/1744-9081-8-33

de Salis, C. A., Rowley, A., Stokell, K., & Brundrett, M. (2019). Do we need more male primary teachers? Tensions and contradictions in the perspectives of male and female trainees. *Education 3–13, 47*(4), 475–489. https://doi.org/10.1080/03004279.2018.1498997.

Dew, K. H., Galassi, J. P., & Galassi, M. D. (1984). Math anxiety: Relation with situational test anxiety, performance, physiological arousal, and math avoidance behavior. *Journal of Counseling Psychology, 31*(4), 580–583. https://doi.org/10.1037/0022-0167.31.4.580

Dickson, M., & Le Roux, J. (2013). Feminising Schools in the Emirate of Abu Dhabi, UAE: Cause for Concern! *Journal of Educational and Psychological Studies, 7*(4), 525–536. https://doi.org/10.24200/jeps.vol7iss4pp525-536.

Dossel, S. (2016). Maths anxiety. *Australian Mathematics Teacher, 72*(3), 40–44.

Dowker, A., Sarkar, A., & Yen Looi, C. (2016). Mathematics anxiety: What have we learned in 60 years? *Frontiers in Psychology, 7*, 1–16. https://doi.org/10.3389/fpsyg.2016.00508

Dreger, R. M., & Aiken, L. R. (1957). The identification of number anxiety in a college population. *Journal of Educational Psychology, 48*(6), 344–351. https://doi.org/10.1037/h0045894

Else-Quest, N. M., Hyde, J. S., & Linn, M. C. (2010). Cross-national patterns of gender differences in mathematics: A meta-analysis. *Psychological Bulletin, 136*(1), 103–127. https://doi.org/10.1037/a0018053

Goetz, T., & Hall, N. C. (2013). Emotion and achievement in the classroom. In J. A. Hattie, E. M. (Ed.), *International guide to student achievement* (pp. 192–195). Routledge.

Goetz, T., Pekrun, R., Hall, N., & Haag, L. (2006). Academic emotions from a social-cognitive perspective: Antecedents and domain specificity of students' affect in the context of Latin instruction. *The British Journal of Educational Psychology, 76*(2), 289–308. https://doi.org/10.1348/000709905X42860

Gough, M. F. (1954). Why failures in mathematics? Mathemaphobia: Causes and treatments. *The Clearing House, 28*(5), 290–294. https://doi.org/10.1080/00098655.1954.11476830

Gresham, G. (2007). A study of mathematics anxiety in pre-service teachers. *Early Childhood Education Journal, 35*(2), 181–188. https://doi.org/10.1007/s10643-007-0174-7

Gresham, G. (2008). Mathematics anxiety and mathematics teacher efficacy in elementary pre-service teachers. *Teaching Education, 19*(3), 171–184. https://doi.org/10.1080/10476210802250133

Gresham, G. (2018). Preservice to inservice: Does mathematics anxiety change with teaching experience? *Journal of Teacher Education, 69*(1), 90–107. https://doi.org/10.1177/0022487117702580

Faust, M. (1992). *Analysis of physiological reactivity in mathematics anxiety (Doctoral dissertation).* Bowling Green State University.

Fennema, E., & Sherman, J. A. (1976). Fennema-Sherman mathematics attitudes scales: Instruments designed to measure attitudes toward the learning of mathematics by females and males. *Journal for Research in Mathematics Education, 7*(5), 324–326. https://doi.org/10.2307/748467

Fraser, B. J., & Fisher, D. L. (1982). Effects of classroom psychosocial environment on student learning. *British Journal of Educational Psychology, 52*(3), 374–377. https://doi.org/10.1111/j.2044-8279.1982.tb02525.x

Freedman, E. (2003). Professor freedman's math help website. http://www.mathpower.com/.

Frenzel, A. C., Pekrun, R., & Goetz, T. (2007). Perceived learning environment and students' emotional experiences: A multilevel analysis of mathematics classrooms. *Learning and Instruction, 17*(5), 478–493. https://doi.org/10.1016/j.learninstruc.2007.09.001

Haciomeroglu, G. (2014). Elementary pre-service teachers' mathematics anxiety and mathematics teaching anxiety. *International Journal for Mathematics Teaching and Learning, 15*(1), 1–10.

Hadfield, O. D., & Lillibridge, F. (1991). *A hands-on approach to the improvement of rural elementary teacher confidence in science and mathematics.* In: *Paper presented at the Annual National Rural Small Schools Conference, Nashville, TN*.

Hadfield, O. D., & McNeil, K. (1994). The relationship between Myers-Briggs Personality Type and mathematics anxiety among preservice elementary teachers. *Journal of Instructional Psychology, 21*(4), 375–384.

Hadley, K. M., & Dorward, J. (2011). Investigating the relationship between elementary teacher mathematics anxiety, mathematics instructional practices, and student mathematics achievement. *Journal of Curriculum and Instruction, 5*(2), 27–44.

Harari, R. R., Vukovic, R. K., & Bailey, S. P. (2013). Mathematics anxiety in young children: An exploratory study. *Journal of Experiential Education, 81*, 538–555. https://doi.org/10.1080/00220973.2012.727888.

Harper, N. W., & Daane, C. J. (1998). Causes and reduction of math anxiety in preservice elementary teachers. *Action in Teacher Education, 19*(4), 29–38. https://doi.org/10.1080/01626620.1998.10462889

Helmke, A. (1983). *Schulische leistungsangst [Scholastic fear of achievement].* Lang Verlag.

Hembree, R. (1988). Correlates, causes, effects, and treatment of test anxiety. *Review of Educational Research, 58*(1), 47–77.

Hembree, R. (1990). The nature, effects, and relief of mathematics anxiety. *Journal for Research in Mathematics Education, 21*(1), 33–46. https://doi.org/10.2307/749455

Hemmings, B. C. (2015). Strengthening the teaching self-efficacy of early career academics. *Issues in Educational Research, 25*(1), 1–17.

Herts, J., & Beilock, S. (2017). From Janet T. Spence's manifest anxiety scale to the present day: Exploring math anxiety and its relation to math achievement. *A Journal of Research, 77*(11), 718–724. https://doi.org/10.1007/s11199-017-0845-9.

Hilton, P. (1980). Math anxiety: Some suggested causes and cures: Part 1. *The Two-Year College Mathematics Journal, 11*(3), 174–188. https://doi.org/10.2307/3026833

Hodges, H., & L. B. (1983). Learning styles: Rx for mathophobia. *The Arithmetic Teacher, 30*(7), 17–20.

Hunt, G. E. (1985). Maths anxiety—Where do we go from here? *Focus on Learning Problems in Mathematics, 7*(2), 29–40.

Jackson, C. D., & Leffingwell, R. J. (1999). The role of instructors in creating math anxiety in students from kindergarten through college. *The Mathematics Teacher, 92*(2), 583–586.

Jenßen, L., Dunekacke, S., Eid, M., & Blömeke, S. (2015). The relationship of mathematical competence and mathematics anxiety: An application of latent state-trait theory. *Zeitschrift Für Psychologie, 223*(1), 31–38. https://doi.org/10.1027/2151-2604/a000197

Kelly, W. P., & Tomhave, W. K. (1985). A study of math anxiety/math avoidance in preservice elementary teachers. *The Arithmetic Teacher, 32*(5), 51–53.

Khasawneh, E., Gosling, C., & Williams, B. (2021). What impact does maths anxiety have on university students? *BMC Psychology, 9*(37), 2–9. https://doi.org/10.1186/s40359-021-00537-2

Kippels, S., & Ridge, N. (2019). The growth and transformation of K–12 education in the UAE. In *Education in the United Arab Emirates* (pp. 37–55). Springer. https://doi.org/10.1007/978-981-13-7736-5_3.

Klieme, E., Hartig, J., & Rauch, D. (2008). The concept of competence in educational contexts. In J. Hartig, Klieme, E., & Leutner, D. (Ed.), *Assessment of competencies in educational contexts: State of the art and future prospects* (pp. 3–22). Hogrefe.

Kogelman, S., & Warren, J. (1978). *Mind over math.* McGraw Hill.

Langton, J. (2017). UAE President announces plans to unify education in all government schools. *The National.* https://www.thenational.ae/uae/uae-president-announces-plans-to-unify-education-in-all-government-schools-1.625336.

Lazarus, M. (1974). Mathophobia: Some personal speculations. *National Elementary Principal, 53,* 16–22.

Levine, G. (1993). Prior mathematics history, anticipated mathematics teaching style, and anxiety for teaching mathematics among pre-service elementary school teachers. In *Paper presented at the North American chapter of the International Groups for the Psychology of Mathematics Education, Assisi.*

Levine, G. (1996). Variability in anxiety for teaching mathematics among pre-service elementary school teachers enrolled in a mathematics course. In *Paper presented at the American Educational Research Association, New York, NY.*

Leso, T. P., & K.L. (1992). Computer anxiety and different types of computer courses. *Journal of Educational Computing Research, 8*(4), 460–478.

Liebert, R. M., & Liebert, L. L. (1998). *Personality strategies and issues.* Brooks.

Liebert, R. M., & Morris, L. W. (1967). Cognitive and emotional components of test anxiety: A distinction and some initial reports. *Psychological Reports, 20,* 975–978.

Lyons, I. M., & Beilock, S. L. (2012). Mathematics anxiety: Separating the math from the anxiety. *Cerebral Cortex, 22,* 2102–2110. https://doi.org/10.1093/cercor/bhr28

Ma, X. (1999). A meta-analysis of the relationship between anxiety toward mathematics and achievement in mathematics. *Journal for Research in Mathematics Education, 30*(5), 520–540. https://doi.org/10.2307/749772

Ma, X., & Xu, J. (2004). The causal ordering of mathematics anxiety and mathematics achievement: A longitudinal panel analysis. *Journal of Adolescence, 27*(2), 165–179. https://doi.org/10.1016/j.adolescence.2003.11.003

Mack, J., & Walsh, B. (2013). Mathematics and science combinations NSW HSC 2001–2011 by gender. Technical paper. http://www.maths.usyd.edu.au/u/SMS/MWW2013.pdf.

Mattarella-Micke, A., Mateo, J., Kozak, M. N., Foster, K., & Beilock, S. (2011). Choke or thrive? The relation between salivary cortisol and math performance depends on individual differences in working memory-anxiety. *Emotion, 11,* 1000–1005. https://doi.org/10.1037/a0023224

McMinn, M. (2019). *Investigating pre-service teachers' mathematics anxiety, teaching anxiety, self-efficacy, beliefs about mathematics and perceptions of the learning environment.* [Unpublished doctoral dissertation]. Curtin University.

McMinn, M., & Aldridge, J. (2020). Learning environment and anxiety for learning and teaching mathematics among preservice teachers. *Learning Environment Research, 23,* 331–345. https://doi.org/10.1007/s10984-019-09304-y

Miller, L. D., & Mitchell, C. E. (1994). Mathematics anxiety and alternative methods of evaluation. *Journal of Instructional Psychology, 21*(4), 353–358.

Morris, J. (1981). Math anxiety: Teaching to avoid it. *The Mathematics Teacher, 74*(6), 413–417.

Muijs, D., & Reynolds, D. (2015). Teache's' beliefs and behaviors: What really matters? *The Journal of Classroom Interaction, 50*(1), 25–40.

Muis, K. R. (2004). Personal epistemology and mathematics: A critical review and synthesis of research. *Review of Educational Research, 74*(3), 317–377. https://doi.org/10.3102/00346543074003317

Naveh-Benjamin, M., McKeachie, W. J., & Lin, Y. G. (1987). Two types of test anxious students: Support for an information processing model. *Journal of Educational Psychology, 79*, 131–136.

Novak, E., & Tassell, J. L. (2017). Studying preservice teacher math anxiety and mathematics performance in geometry, word, and non-word problem solving. *Learning and Individual Differences, 54*, 20–29. https://doi.org/10.1016/j.lindif.2017.01.005

Núñez-Peña, M. I., & Suárez-Pellicioni, M. (2015). Processing of multi-digit additions in high math-anxious individuals: Psychophysiological evidence. *Frontiers in Psychology, 6*(1268), 1–11. https://doi.org/10.3389/fpsyg.2015.01268

Nurlu, Ö. (2015). Investigation of teache's' mathematics teaching self-efficacy. *International Electronic Journal of Elementary Education, 8*(1), 21–40.

Okur, M., & Bahar, H. H. (2010). Learning styles of primary education prospective mathematics teachers; states of trait-anxiety and academic success. *Procedings of Social and Behavioral Sciences, 2*(2), 3632–3637. https://doi.org/10.1016/j.sbspro.2010.03.565.

Park, G. P., & French, B. F. (2013). Gender differences in the foreign language classroom anxiety scale. *System, 41*, 462–471. https://doi.org/10.1016/j.system.2013.04.001

Peker, M., & Ertekin, E. (2011). The relationship between mathematics teaching anxiety and mathematics anxiety. *The New Educational Review, 23*(1), 213–226.

Peker, M., & Ulu, M. (2018). The effect of pre-service mathematics teache's' beliefs about mathematics teaching-learning on their mathematics teaching anxiety. *International Journal of Instruction, 11*(3), 249–264. https://doi.org/10.12973/iji.2018.11318a

Plake, B. S., & Parker, C. S. (1982). The development and validation of a revised version of the mathematics anxiety rating scale. *Educational and Psychological Measurement, 42*(2), 551–557. https://doi.org/10.1177/001316448204200218

Pletzer, B., Kronbichler, M., Nuerk, H. C., & Kerschbaum, H. H. (2015). Mathematics anxiety reduces default mode network deactivation in response to numerical tasks. *Frontiers in Human Neuroscience, 9*, 202. https://doi.org/10.3389/fnhum.2015.00202

Pletzer, B., Wood, G., Moeller, K., Nuerk, H. C., & Kerschbaum, H. H. (2010). Predictors of performance in a real-life statistics examination depend on the individual cortisol profile. *Biological Psychology, 85*, 410–416. https://doi.org/10.1016/j.biopsycho.2010.08.015

Pourmoslemi, A., Erfani, N., & Firoozfar, I. (2013). Mathematics anxiety, mathematics performance and gender differences among undergraduate students. *International Journal of Scientific and Research Publications, 3*(7), 1–6.

Ramirez, G., & Beilock, S. L. (2011). Writing about testing worries boosts exam performance in the classroom. *Science, 331*, 211–213. https://doi.org/10.1126/science.1199427

Ramirez, G., Chang, H., Maloney, E. A., Levine, S. C., & Beilock, S. L. (2016). On the relationship between math anxiety and math achievement in early elementary school: The role of problem solving strategies. *Journal of Experimental Child Psychology, 141*(C), 83–100. https://doi.org/10.1016/j.jecp.2015.07.014.

Ramirez, G., Gunderson, E. A., Levine, S. C., & Beilock, S. L. (2013). Math anxiety, working memory, and math achievement in early elementary school. *Journal of Cognition and Development, 14*(2), 187–202. https://doi.org/10.1080/15248372.2012.664593

Reilly, D., Neumann, D. L., & Andrews, G. (2019). Investigating gender differences in mathematics and science: Results from the 2011 trends in mathematics and science survey. *Research in Science Education, 49*(1), 25–50.

Richardson, F. C., & Suinn, R. M. (1972). The mathematics anxiety rating scale: Psychometric data. *Journal of Counseling Psychology, 19*(6), 551–554. https://doi.org/10.1037/h0033456

Richardson, F. C., & Woolfolk, R. L. (1980). Mathematics anxiety. In I. G. Sarason (Ed.), *Test anxiety: Theory, research, and applications* (pp. 271–288). Erlbaum.

Roos, A. L., Bieg, M., Goetz, T., Frenzel, A. C., Taxer, J., & Zeidner, M. (2015). Experiencing more mathematics anxiety than expected? Contrasting trait and state anxiety in high achieving students. *High Ability Studies, 26*(2), 245–258. https://doi.org/10.1080/13598139.2015.1095078

Rounds, J. B., & Hendel, D. D. (1980). Measurement and dimensionality of mathematics anxiety. *Journal of Counseling Psychology, 27*(2), 138–149. https://doi.org/10.1037/0022-0167.27.2.138

Sandman, R. S. (1980). The mathematics attitude inventory: Instrument and user's manual. *Journal for Research in Mathematics Education, 11*(2), 148–149. https://doi.org/10.2307/748906

Schmidt, W. H., & Buchmann, M. (1983). Six teachers' beliefs and attitudes and their curricular time allocations. *The Elementary School Journal, 84*(2), 162–171. https://doi.org/10.1086/461353

Sloan, T. (2010). A quantitative and qualitative study of math anxiety among preservice teachers. *The Educational Forum, 74*(3), 242–256. https://doi.org/10.1080/00131725.2010.483909

Sommer, M., & Arendasy, M. E. (2016). Does trait test anxiety compromise the measurement fairness of high-stakes scholastic achievement tests? *Learning and Individual Differences, 50*, 1–10.

Spielberger, C. D., & Vagg, R. P. (1995). Test anxiety: A transactional process model. In C. D. Spielberger & P. R. Vagg (Eds.), *Test anxiety: Theory* (pp. 3–14). Taylor & Francis.

Spicer, J. (2004). Resources to combat math anxiety. *Eisenhower National Clearinghouse Focus, 12*(12), 1–8.

Stoehr, K. J. (2017). Mathematics anxiety: One size does not fit all. *Journal of Teacher Education, 68*(1), 69–84. https://doi.org/10.1177/0022487116676316

Suinn, R. M., & Winston, E. H. (2003). The mathematics anxiety rating scale, a brief version: Psychometric data. *Psychological Reports, 92*(1), 167–173. https://doi.org/10.2466/pr0.2003.92.1.167

Suman, O. O., Caglayan, K. Y., & Kartal, A. (2015). Fear of mathematics of pre-service primary school teachers. *H. U. Journal of Education, 30*(2), 69–80.

Swetman, B., Munday, R., & Windham, R. (1993). Math-anxious teachers: Breaking the cycle. *College Student Journal, 22*(4), 421–427.

Taylor, J. A. (1953). A personality scale of manifest anxiety. *Journal of Abnormal Psychology, 48*, 285–290.

Taylor, L. & Brooks, K. (1986). Building math confidence by overcoming math anxiety. From theory to practice. *Adult Literacy and Basic Education, 10*(1), 2–5.

Taylor, B. A., & Fraser, B. J. (2013). Relationships between learning environment and mathematics anxiety. *Learning Environments Research, 16*(2), 297–313. https://doi.org/10.1007/s10984-013-9134-x

The Abu Dhabi Government. (2008). *The Abu Dhabi Economic Vision Report 2030.* Abu Dhabi, UAE.

Tobias, S. (1980). *Overcoming math anxiety.* Houghton Mifflin.

Tobias, S. (1985). Test anxiety: Interference, defective skills, and cognitive capacity. *Educational Psychologist, 20*, 135–142.

Tobias, S. (1990). Math anxiety: An update. *NACADA Journal, 10*(1), 47–50.

Tocci, C. M., & Engelhard, G. (1991). Achievement, parental support, and gender differences in attitudes toward mathematics. *The Journal of Educational Research, 84*(5), 280–286. https://doi.org/10.1080/00220671.1991.10886028

Tooke, D. J., & Lindstrom, L. C. (1998). Effectiveness of a mathematics methods course in reducing math anxiety of preservice elementary teachers. *School Science and Mathematics, 98*(3), 136–140.

Trujillo, K. M., & Hadfield, O. D. (1999). Tracing the roots of mathematics anxiety through in-depth interviews with preservice elementary teachers. *College Student Journal, 33*(2), 219.

Uusimaki, L., & Nason, R. (2004, July). Causes underlying pre-service teachers' negative beliefs and anxieties about mathematics. In *Paper presented at the 28th Conference of the International Group for the Psychology of Mathematics Education, Bergen, Norway.*

Vinson, B. M. (2001). A comparison of preservice teache's' mathematics anxiety before and after a methods class emphasizing manipulatives. *Early Childhood Education Journal, 29*(2), 89–94. https://doi.org/10.1023/A:1012568711257

von der Embsea, N., Jesterb, D., Royb, D., & Postb, J. (2018). Test anxiety effects, predictors, and correlates: A 30-year meta-analytic review. *Journal of Affective Disorders, 227*, 483–493.

Walsh, J. J., & Ugumba-Agwunobi, G. (2002). Individual differences in statistics anxiety: The roles of perfectionism, procrastination and trait anxiety. *Personality and Individual Differences, 33*(2), 239–251. https://doi.org/10.1016/S0191-8869(01)00148-9

Whyte, J., & Anthony, G. (2012). Maths anxiety: The fear factor in the mathematics classroom. *New Zealand Journal of Teachers' Work, 9*(1), 6–15.

Widmer, C. C., & Chavez, A. (1982). Math anxiety and elementary school teachers. *Education, 102*(3), 272–276.

Wigfield, A., & Meece, J. L. (1988). Math anxiety in elementary and secondary school students. *Journal of Educational Psychology, 80*(2), 210–216. https://doi.org/10.1037/0022-0663.80.2.210

Wine, J. (1971). Test anxiety and direction of attention. *Psychological Bulletin, 76*, 92–104. https://doi.org/10.1037/h0031332.

Wittmaier, B. (1972). Test anxiety and study habits. *Journal of Educational Research, 65*, 852–854.

Wood, E. F. (1988). Math anxiety and elementary teachers: What does research tell us? *For the Learning of Mathematics, 8*(1), 8–13.

Young, C. B., Wu, S. S., & Menon, V. (2012). The neurodevelopmental basis of math anxiety. *Psychological Science, 23*, 492–501. https://doi.org/10.1177/0956797611429134

Melissa McMinn has worked in in-service and pre-service teacher education for 15 years and in post-graduate education for eight years. She is currently facilitating a new suite of initial education teacher programs for the Open Polytechnic in New Zealand. She holds a Master of Education and a Doctorate of Philosophy in Mathematics and Science Education and has achieved the status of Senior Fellow (SFHEA) in recognition of her teaching and learning support in higher education. She is an active researcher and over the past decade has led projects in mathematics and science anxiety in university students and teachers and learning environments research, and co-researched a number of projects including pedagogy in higher education and children's perceptions of science and technology, among others.

Chapter 6
Gender Differences in ICT Literacy: ICT-Related Individual Characteristics and Enabling Factors

Ieda M. Santos and Shaljan Areepattamannil

Abstract Information and Communication Technology (ICT) literacy is an essential competence that adolescents must possess to successfully participate in the twenty-first-century digital society and workforce. ICT literacy is also crucial to Science, Technology, Engineering and Mathematics (STEM) fields, and it can significantly influence females' STEM-related career paths. This chapter provides background literature and discusses key ICT-related individual characteristics predicting gender differences in ICT literacy. It also discusses the main enabling factors and initiatives implemented by the United Arab Emirates (UAE) to support equal opportunities for adolescents to develop ICT literacy skills. The chapter includes recommendations for practice and future research.

Keywords ICT Literacy · Gender differences · STEM · Secondary students · Educational policies

Introduction

Rapid developments in information and communication technologies (ICT) are changing the skill sets required for individuals to successfully live and work in the globalised digital society (Bejaković & Mrnjavac, 2020; World Economic Forum, 2016). ICT has become more relevant in adolescents' lives (Christoph et al., 2015), and their access to technology has increased dramatically in the past few years (OECD, 2015). Thus, ICT literacy represents essential knowledge and skills for adolescents to thrive in the digital society and workplace (Gnambs, 2021). Discrepancies in ICT literacy, such as lack of self-confidence in using computers and glaring differences in the socioeconomic backgrounds, may lead to inevitable socioeconomic inequalities (Senkbeil et al., 2013; van Dijk, 2006). ICT literacy is also crucial to STEM (Science, Technology, Engineering, and Mathematics) as these two concepts complement each other. The latter provides a platform for students to apply their ICT

I. M. Santos (✉) · S. Areepattamannil
Emirates College for Advanced Education, Abu Dhabi, UAE
e-mail: isantos@ecae.ac.ae

M. Dickson et al. (eds.), *Gender in STEM Education in the Arab Gulf Countries*,
https://doi.org/10.1007/978-981-19-9135-6_6

literacy skills "in designing technological devices and solutions—such as robotics and app design—and solving complex problems" (UNICEF, 2020, p. 9). Furthermore, there is growing evidence that 98% of STEM-related jobs would require ICT skills in some form or another (Chavatzia, 2017).

Among the factors that may influence STEM orientation (e.g., self-efficacy, school context, teacher, and parent role model), ICT literacy skills can significantly impact females' STEM-based career intentions (Chavatzia, 2017; Iroaganachi et al., 2020). Iroaganachi et al. (2020, p.12), for example, concluded that the extent to which females perceive they possess ICT literacy skills and use them to "access any information and perform any activity required to make decisions" would sustain their intentions to pursue STEM fields. Although the focus of this chapter is on ICT literacy, it is worth noting, that it is essential to situate ICT literacy within a broader social context that considers other challenging factors influencing females to choose or not to choose STEM careers. Research, policymakers and other stakeholders need to address these factors.

Studies investigating gender differences in ICT literacy among adolescents have shown that the gender gap is narrowing (Siddiq & Scherer, 2019) or, in some cases, females outperform males (Hohlfeld et al., 2013). However, gender differences still exist worldwide where females underachieve in ICT-related activities including holding less positive attitudes towards ICT (Cai et al., 2017; Gnambs, 2021) and showing lower ICT self-efficacy and interest in using computers (Fraillon et al., 2014; Scherer & Siddiq, 2015) as compared to males. Given that ICT literacy is a crucial factor influencing STEM orientation, this chapter discusses key ICT-related behavioural and motivational characteristics predicting gender differences and factors that might contribute to equal opportunities for adolescents to develop ICT literacy skills. The discussions are informed by the International Computer and Information Literacy (ICILS) results (Fraillon et al., 2014, 2019; Gebhardt et al., 2019) and supported by relevant literature on these factors. The authors expect to generate rich discussions and recommendations both in terms of research areas and policymaking that benefit stakeholders in the United Arab Emirates (UAE) since the gender gap persists in STEM-related career aspirations (Patterson et al., 2021).

The chapter first provides background by discussing the concept of ICT literacy and then discusses ICT literacy gender differences. The following sections present ICT-related factors predicting gender differences in ICT literacy and enabling factors. The latter section also considers the reforms and initiatives in the UAE context. The chapter ends with concluding remarks and recommendations for practice and future research.

The Concept of ICT Literacy

Several terminologies have been proposed to address the skills and knowledge needed to critically process technology resources in the digital era (van Laar et al., 2017), including digital competencies (Ferrari, 2013), digital literacy (Roblyer & Hughes,

2019), and ICT literacy (ETS, 2002). ICT literacy is usually defined from a practical perspective (Gnambs, 2021) as individuals' ability to effectively use technology and communication tools to access, manage, integrate, evaluate and create information (ETS, 2002). Both the Organisation for Economic Co-operation and Development (OECD) (Lennon et al., 2003) and ICILS (Fraillon et al., 2013) cover similar areas of ICT literacy in their definitions. Fraillon et al., (2013, p. 17), for instance, refer to ICT literacy as "an individual's ability to use computers to investigate, create, and communicate in order to participate effectively at home, at school, in the workplace, and in society." In this case, ICT literacy is conceptualised based on four main strands, namely 1) understanding computer usage, 2) collecting, 3) creating and 4) communicating information. However, ICT literacy must be seen as an evolving concept due to technological advancements as stressed by UNESCO (2018, p. 2):

> a person mastering digital skills at a given time is likely to lose his or her mastery because of technological evolution. This dimension must be taken into account when grasping the subject of digital skills: those skills must be not only acquired but also constantly adapted and updated. For example, rapid development of artificial intelligence and the "Internet of Things" will give rise to new skills…

The above definitions share a common understanding that ICT literacy is an essential competence that individuals must possess to "participate effectively and flourish in a modern society, in terms of economic and psychological well-being" (Gnambs, 2021, p. 2). These definitions assume that individuals are knowledgeable about technical skills to efficiently use digital technologies and identify similar sets of integrated processes (Fraillon et al., 2015) presented in a sequence, implying increased cognitive difficulty (ETS, 2002). Moreover, the skills and knowledge identified in these definitions are relevant to STEM fields (Chavatzia, 2017) and its advancements in rapidly emerging technological developments, including artificial intelligence and the Internet of things (UNESCO, 2019). In this sense, ICT literacy complements STEM areas where the former is employed in STEM activities to solve problems (Dogan & Robin, 2015) such as climate change. Therefore, as ICT is a fundamental literacy for everyday life, work and to support STEM fields, policies must ensure equal opportunities for all students, irrespective of grade level, race, ethnicity, cultural background, socioeconomic status and gender, to develop the needed skills for them to excel in the future (Borokhovski et al., 2018).

Gender Differences in ICT Literacy

Despite the emphasis on equal gender opportunities for all in the knowledge society, a significant number of studies have found individual differences in ICT literacy among adolescents favouring males (ACARA, 2012; Fraillon et al., 2014; Kuhlemeier & Hemker, 2007). For example, Kuhlemeier and Hemker (2007) found that males showed more developed Internet skills than females. These studies confirm the general understanding that males hold more positive perceptions towards technology use, thus performing better than females. However, other findings showed opposite results where female students scored significantly higher than males in ICT literacy skills (Fraillon et al., 2019; Hohlfeld et al., 2013; Ritzhaupt et al., 2013). Based on data from ICILS, Gebhardt et al. (2019), for instance, found that females outperformed males in most countries. However, a more in-depth investigation based on differential item functioning (DIF) analysis revealed that females performed better on creating information than males. In contrast, males outperformed females on technical tasks. Similarly, Punter et al. (2017) reported that girls outperformed boys in all ICILS countries on sub-scales related to sharing or communicating information and evaluating and reflecting on the information.

Further research suggests that the gender gap in ICT literacy is narrowing (Gnambs, 2021; Siddiq & Scherer, 2019) or non-existent (Claro et al., 2012; Jan, 2018; van Deursen & van Diepen, 2013). For example, Siddiq and Scherer's (2019) meta-analysis, comprising 46 studies, indicated slight gender differences in ICT literacy favouring females. In the same vein, van Deursen and van Diepen (2013) found no differences in information and strategic skills to use the Internet among Dutch secondary boys and girls of various age groups. Increased access to ICT might be contributing to reducing the gender gap in ICT skills (Borokhovski et al., 2018), although results by Cai et al. (2017) do not support this statement.

Despite the increased access to technologies and their integration in the school curriculum, Cai et al. (2017) found that females, including secondary level, still hold a less positive attitude towards ICT use. This result supports Punter et al.'s (2017) argument that it is premature to conclude that the ICT gender gap has closed without further research. This observation is relevant as gender differences concerning individual characteristics, such as ICT self-efficacy tend to widen or close the gender gap. In addition, and according to Cussó-Calabuig et al. (2018), when looking at these specific factors regarding ICT use, gender differences can become more evident.

ICT-Related Individual Characteristics Predicting Gender Differences in ICT Literacy

ICT-related individual characteristics, such as use and engagement with ICT at school and home (e.g., types of computer applications and frequency of use), and ICT self-efficacy, are essential pre-conditions to develop ICT literacy (Claro et al., 2012; Fraillon et al., 2019). Such information can be helpful for "formulating supportive policies and opportunities" (Cai et al., 2017, p. 10) to address gender gaps among adolescents and support their development of ICT literacy skills. The following sections discuss key behavioural and motivation characteristics that can contribute to the acquisition of ICT literacy.

Behavioural Characteristics

Research has shown that ICT-related behaviour characteristics, including students' access, use and engagement with computer-based applications at school or home, can positively impact ICT literacy (Aesaert et al., 2015; Claro et al., 2012; Fraillon et al., 2019; Hatlevik et al., 2018). In this context, the use of technologies offers adolescents opportunities to acquire essential skills required both at work and in life. Thus, gender differences in ICT access and use may lead to differences concerning technologies and how one approaches future careers (Cussó-Calabuig et al., 2018).

Out-of-School ICT Activities

Based on data from PISA 2006, Drabowicz (2014) found that male students used computers at home more often than females in most of the countries under study (39 countries). In most of these countries, males also reported using computers at locations other than home or school more often than females. Further studies found that males used ICT more frequently outside school for leisure than females, particularly for games and music (Colley & Comber, 2003; Fairlie, 2015; Lim & Meier, 2011; OECD, 2019; Zhong, 2011). The ICILS found similar results; however, in some countries, females reported more frequent use of computers for leisure (Fraillon et al., 2019; Gebhardt et al., 2019). Other studies showed that females reported using computers more often for social communication purposes (Cussó-Calabuig, 2018; Fairlie, 2015), such as extensive use of online social networking websites (Lim & Meier, 2011).

Similarly, Gebhardt et al. (2019) found that female students used social communication in most ICILS countries more frequently. However, no gender difference was found in some countries, whereas in Turkey, males used social media more frequently, such as voice chat. Drabowicz (2014) revealed mixed results where males reported using computers more often for communication purposes than females in ten out of the 39 countries studied, while in 19 countries, the results were reversed. In the remaining countries, the study did not find gender differences.

Drabowicz (2014) challenges the above studies that suggested a clear pattern where males and females differ in their gaming and social networking activities. Nevertheless, non-school ICT-related activities may facilitate familiarity with technologies and create opportunities for informal learning that develop digital skills (Gebhardt et al., 2019; OECD, 2019). Research suggested that adolescents' ICT access and use at home are positively associated with perceived ICT skills (Cussó-Calabuig et al., 2018; Zhong, 2011). Kuhlemeier and Hemker (2007), for example, found that home access and computer usage to perform activities such as web surfing and chatting were significantly associated with the development of Internet and computer skills. In this study, males showed more developed Internet skills than females.

Although out-of-school computer activities may negatively impact academic achievement (Drabowicz, 2014), playing games, for example, may help develop visual-spatial skills (Jaeggi et al., 2014). The development of these skills can directly impact future STEM career paths. According to Lubinski (2010, p. 348), "the consistently distinguished levels of spatial ability among adolescents who subsequently go onto earn STEM educational degrees and occupations … reveals the importance of spatial ability in STEM arenas." Along these lines, the fact that male students are playing games more often than females might decrease interest in pursuing STEM by the latter. Research could investigate the relationship between playing games and females' STEM career orientation or selection of STEM-related degrees.

School-Related ICT Activities

Studies have suggested that female students are more likely than males to use computers for school-related activities (e.g., ACARA, 2014; Fraillon et al., 2019). Similarly, other results revealed that female students reported more frequent use of home computers for school-related activities than males, although the difference between gender was minimal (Cussó-Calabuig et al., 2018; OECD, 2011). On the other hand, on average across the OECD countries, more males than females reported using school computers frequently (OECD, 2011), except in a few countries where results were reversed with more females than males reporting frequent use of computers at school. According to BECTA (2008), these differences may be due to males spending more time at school playing computer games outside class time rather than differences in use during class. In the same vein, Drabowicz (2014) found that male students used computers and the Internet more often than females for school-related activities. However, according to Drabowicz, these results may be due to males using more educational software and Internet out of educational necessity rather than wanting to use it as another way for improving their already good academic achievement. However, this interpretation needs to be validated with further research.

Motivational Characteristics

Research has also emphasised the importance of ICT-related motivational factors for adolescents' ICT literacy (Christoph et al., 2015; Fraillon et al., 2014; Hatlevik et al., 2018; Senkbeil, 2021). Among motivational constructs, self-efficacy, interest and enjoyment have been identified as key predictors of ICT literacy (Christoph et al., 2015; Fraillon et al., 2014; Senkbeil, 2021). These main constructs are discussed in the next sections.

ICT Self-efficacy

Perceived self-efficacy refers to an individual's beliefs in performing tasks with those abilities they possess (Bandura, 1986). When individuals believe they can perform tasks, they become motivated to accomplish the goal (Tømte & Hatlevik, 2011). In contrast, individuals with low confidence in their ability to use technology are considered to have low self-efficacy beliefs (Eastin & LaRose, 2006). Therefore, how girls perceive self-efficacy of ICT skills plays a critical role in choosing STEM path careers (Chavatzia, 2017).

According to Fraillon et al. (2014), ICT self-efficacy can be classified into basic and advanced ICT skills. The same authors found that females reported slightly higher levels of ICT self-efficacy than males on basic ICT activities (common ICT applications). In contrast, males reported much higher levels of ICT self-efficacy concerning more complex ICT activities (multimedia and technical aspects). Although focusing on basic ICT skills only, Hatlevik et al. (2018) reported similar results in favour of females. In contrast, Cussó-Calabuig et al. (2018) found no differences in self-efficacy between males and females in Catalonia. However, when analysing specific tasks, their findings corroborate the ICILS' results (Fraillon et al., 2014), where girls reported higher self-efficacy than boys in basic tasks and boys reported higher efficacy in advanced tasks than girls. Other studies presented mixed results regarding ICT self-efficacy and gender differences (Gebhardt et al., 2019; Tømte & Hatlevik, 2011; Tsai & Tsai, 2010). For example, Gebhardt et al. (2019) indicated that females reported significantly higher levels of self-efficacy in ICT general skills in six countries in ICILS 2013 while in the remaining countries, gender difference was either small or non-existent.

These findings suggest that female self-efficacy in more complex ICT skills is significantly lower than males overall. However, researchers (Gnambs, 2021; Punter et al., 2017) cautioned that ICT self-efficacy differences between males and females maybe due to the former overestimating their ICT abilities while the latter underestimating their skills. According to Zhong (2011), one possible explanation for this pattern may be that males use more computers from home and engage in self-learning, which enhances their ICT skills. ICT home access and use might explain males' self-confidence in using technologies. Also, the stereotypical notion that computer is a male domain may lead to females' perceived lack of self-efficacy in ICT (BECTA, 2008).

Interest and Enjoyment

Interest and enjoyment are also important concepts predicting ICT literacy. Interest is a "psychological state of engaging or the predisposition to reengage with particular classes of objects, events, or ideas over time" (Hidi & Renninger., 2006, p. 112). Interest in ICT emphasises adolescents' positive assessments of ICT and ICT-related activities (Christoph et al., 2015), and enjoyment of their use and recognising related benefits from using ICT (Zylka et al., 2015). Hohlfeld et al. (2013) noted that students must have positive attitudes and interests to develop ICT skills. The more interested in computers adolescents are, the more computer skills they demonstrate (Christoph et al., 2015).

However, when analysing gender differences, males reported greater interest and enjoyment in computers than females (Cussó-Calabuig et al., 2018; Fraillon et al., 2014; Gebhardt et al., 2019), which has influence on the performance of males in ICT (Gebhardt et al., 2019). However, these findings are not consistent with other results where gender difference was non-existent (Christoph et al., 2015; Teo, 2008). Kunina-Habenicht and Goldhammer (2020) also reported similar findings; however, further analysis for Germany and Switzerland PISA 2015 revealed that ICT interest was considerably lower for girls in the German sample. Despite the mixed results, these findings reinforce stereotypes that computer-related interests are male-specific, which broaden the gender gap in student' pathways towards STEM careers.

Key Factors Facilitating ICT Literacy: Teachers, Schools and Policies

This section discusses three main factors that may contribute to (or hinder) equal opportunities for females and males to develop ICT literacy skills. The discussions are based on the results of ICILS studies (Fraillon et al., 2014, 2019; Gebhardt et al., 2019) and relevant literature on the same topic. Although the ICILS results are based on developed countries, they offer valuable information on factors contributing to the development of ICT literacy based on these country members' experiences.

Teachers

Teachers' use of ICT in the classroom is positively associated with students' digital skills development (Fraillon et al., 2014). These results from the ICILS corroborate previous studies showing a correlation between student perceived use of ICT and pedagogies adopted by teachers using ICT (Law et al., 2008). ICT in the classroom is mediated by teachers who integrate an increased range of digital technologies (Davis et al., 2013) within constructivist learner-centred approaches (Vekiri, 2008) to prepare students

> for future participation in an evolving society where new media practices are deeply embedded in the associated structures and processes. As such, digital literacy encompasses the purpose, setting and practices in which technology is used. In this way digital literacy has a special meaning in the classroom – bound up in the relationship between activity and purpose – learning and literacy are interlinked (Casey & Bruce, 2011, p. 3).

The ICILS showed that teachers, who were using ICT extensively for teaching and learning, reported fostering ICT literacy such as searching and presenting information at the lower and upper secondary levels, across most participating countries. These were done through single subjects or integration in other subjects (e.g., science and technology studies) (Fraillon et al., 2019). In this regard, ICT is used as a tool to support teaching and learning, and, at the same time, "recognized as a subject of learning in themselves" (OECD, 2018, p. 45). Within this approach, students can, among other things, learn how to evaluate and synthesise information, tackle real-life problems and foster ICT self-efficacy and interest (Vekiri, 2008).

Of course, one needs to consider that several factors may hinder teachers' effective use of ICT in the classroom, including competencies, self-efficacy, attitudes and age (Eickelmann, 2011; Ertmer, 1999; Gerick et al., 2017). These issues need to be addressed by policies and their application in schools to facilitate effective ICT integration in the classroom. For example, across the ICILS countries, older teachers showed less confidence in using ICT while those who used ICT more often in their teaching were more confident about using technologies (Fraillon et al., 2019). Fraillon et al. (2019) reflected on whether schools should plan targeted programmes to support the development of ICT use by older teachers to help address the imbalance between ages. Alternatively, a more equitable approach could be to develop a community of practice (Wenger, 1998) in the school where all teachers, irrespective of age, experience and subjects, engage in research-informed professional activities to learn with and from peers.

School Learning Environment

Schools also play a crucial role in supporting the integration of ICT in the curriculum (Eickelmann, 2011) and preparing students with the skills required to participate in the global society (Senkbeil, 2021). Schools must invest in professional development for teachers to effectively integrate ICT into the curriculum (UNESCO, 2018). In this respect, visionary school leaders "create a culture where teachers and learners are empowered to use technology in innovative ways to enrich teaching and learning" (International Society for Technology in Education [ISTE], 2016, p. 7). A longitudinal study by Eickelmann (2011) illustrated schools that successfully integrated ICT in the classroom because they had strong leadership, with a sound understanding of the benefits of technologies in the curriculum. Along these lines, Li et al. (2016) found that school leaders' support to innovate using ICT positively influenced their teachers to innovate and use ICT in their teaching. In contrast, as suggested by Eickelmann (2011), leaders who tended to rely on individuals or innovation champions and did not develop a holistic supporting structure for ICT integration, failed to implement ICT sustainably in their schools. In addition, these schools

> have focused on teaching ICT competencies instead of using ICT to enhance learning. For example, they organized courses on using a word processing or spreadsheet software instead of relating their application to content or suggesting an overarching pedagogical approach (Eickelmann, 2011, p. 94).

Furthermore, the availability of school technological infrastructure (e.g., networks, software computers) is a crucial element for developing ICT literacy skills (UNESCO, 2018). Zhong (2011), for example, found a positive association between access to ICT school facilities and student self-reported digital skills. However, the availability of technologies in schools alone is not enough to ensure that students and teachers will use them effectively (Cussó-Calabuig et al., 2018). *How* students use computers is critical for learning and developing digital skills (Fraillon et al., 2019). For example, the ePIRLS results suggested that students' daily use of a computer at school to prepare essays or homework showed higher digital skills levels than others (Mullis et al., 2017). On the other hand, those, who had spent the same amount of time on a computer at school but searching online information only, demonstrated slightly lower digital skills than others. Thus, school policies should provide the conditions where technologies are used supportively and are valuable in the classrooms (Casey & Bruce, 2011).

Educational Policies

Policies are also developed to respond to the dual goal of improving the effectiveness of ICT integration in schools and enhancing students' ICT literacy skills (UNESCO, 2018). According to Kozma (2008), ICT-related policies often cover a) technological infrastructure, b) teacher professional development, c) technical support, d) pedagogical and curricular modification and e) content development. Similarly, the ICILS reported almost full agreement across participating countries that their policies emphasised priorities that included professional development for teachers' pedagogic use of ICT, development of digital learning materials and reduction of the digital divide between groups of students. Regarding the latter, policies must ensure equity among students, including vulnerable groups (UNESCO, 2018), and "Increase education funding to ensure that all students enjoy a parity of devices and online access" (International Literacy Association, 2017, p. 4).

Similar policies are reflected on the Education Technology Master Plan by the Ministry of Education in Singapore (Ministry of Education Singapore, 2021). Their plan consists of several strands, including professional development for teachers to learn how to use and integrate technologies in teaching and provide schools with the necessary technological infrastructure. It is worth noting their emphasis on supporting parents with relevant resources and suggestions to ensure their children have access to technology at home, to learn. This policy is relevant as research has shown that students are also developing ICT digital skills informally outside school contexts (Claro et al., 2012; Fraillon et al., 2019).

Teacher professional development, an essential component of these policies, includes training on how to use ICT and integration of ICT into teaching (UNESCO, 2018). The provision of professional development for teachers on ICT use is positively related to its integration in the classroom (Gerick et al., 2017). Thus, policies must consider the provision of time for teachers to develop skills confidence and learn how to apply technologies in teaching activities (Fraillon et al., 2019). It must also view professional development as lifelong learning to update skills and pedagogies (Tømte, 2011). Finally, meaningful professional development activities should take place within a community of practice that considers teachers' context and experiences (Dlamini & Mbatha, 2018).

UAE Initiatives

In the UAE context, alongside reforms to curriculum and standardisation of Teacher Licensing, there are government efforts in recent years to integrate ICT in schools. These efforts recognised ICT's importance in developing competencies and digital skills to prepare Emirati students for active roles and participation in the digital society (Alaleeli & Alnajjar, 2020; OECD, 2020). Tamim (2013, p. 49) added that "there is a growing awareness for the need to enhance the curriculum with the objective of utilising technology to provide students with a better learning environment."

The Ministry of Education (MOE) has provided funds to improve schools' technological infrastructure while upgrading the skills of teachers by offering ICT training (Ministry of Education UAE, 2021). The introduction of computers and tablets in all schools are clear examples of the government's initiatives to equip schools with technologies to support teaching practices. In addition, the MOE

> has launched numerous initiatives and programs to develop students' skills and capacities in robotics and artificial intelligence such as robotics training camp, VEX Robotics Championship and innovation Ambassadors. It has also created 11 FabLabs across its schools … Innovative smart portals and platforms have also been created by the Ministry to offer students, teachers, principals and parents support materials, online resources to raise the levels and outcomes of teaching and learning, as well as enhancing student interaction and experience (Ministry of Education UAE, 2019, para. 5&6)

Regarding teacher use of ICT in UAE classrooms, small-scale research by Almekhlafi and Almeqdadi (2010) revealed that teachers integrated technologies in their classes with varying degrees and effectiveness despite the many barriers discussed, such as lack of training. More recently, and according to TALIS 2018 data (OECD, 2020, 2021), teachers in the UAE, both at lower (77%) and upper secondary (81%) levels, reported letting their students use ICT for projects or classroom work. In addition, most of the lower secondary teachers (86%) indicated feeling prepared to adopt ICT in their teaching. This positive outcome may be due to most teachers (87%) stating they had learned ICT integration during their initial teacher training and that 85% had recently participated in ICT professional development (OECD, 2020). Still, there is room for improvement as teachers at the upper (9%) and lower secondary levels (10%) reported a high need for ICT professional development (OECD, 2021). On the positive side, 53% of teachers in upper secondary and 54% in lower secondary levels in the UAE reported engaging in collaborative professional development at least once a month (OECD, 2020). Considering that this type of professional learning positively impacts students' development of ICT digital skills (Gerick et al., 2017), overall, only half of UAE teachers in the secondary level benefit from these collaborative activities.

Given the emphasis on preparing UAE nationals for futures jobs in STEM fields and the shift to a more sustainable economic model (Makhmasi et al., 2012), the continuous reforms by the UAE government might support equal opportunities for Emiratis to develop the needed skills and interest to pursue careers in STEM. For example, although conducted in primary public schools, a study by Ali et al. (2019) illustrated how the effective integration of 3D printing in learning activities in school classrooms was positively associated with interest in STEM-related careers among Emirati female students.

Conclusion and Recommendations

ICT literacy is vital for adolescents' successful participation in the knowledge society (Gnambs, 2021). It is essential for STEM fields (Dogan & Robin, 2015) and can influence females' STEM-related career paths (Chavatzia, 2017; Iroaganachi et al., 2020). While several studies suggested that gender differences in ICT literacy skills are narrowing (Siddiq & Scherer, 2019), other findings are inconsistent (e.g., Cai et al., 2017; Drabowicz, 2014) suggesting that the gender gap and stereotypes persist. Accordingly, male adolescents show more confidence, interest and enjoyment with ICT (e.g., Cussó-Calabuig et al., 2018; Fraillon et al., 2014), and use more ICT outside school, especially for playing games (Zhong, 2011) as opposed to females who use more social networking (Cussó-Calabuig, 2018). Possibly, outside class activities and types of activities using technologies based on adolescents' interests may be contributing to gender differences in ICT literacy. For example, computer gaming might help improve ICT abilities such as visual-spatial skills and orientations towards STEM (Jaeggi et al., 2014). Atit et al. (2020) further noted that spatial skills (e.g., mental manipulation and reasoning) are critical for mathematics. This study showed that females had lower scores on spatial skills than male students. On the other hand, social networking might help females develop interest in ICT as noted by Gnambs (2021). However, this application may not be sufficient to develop higher levels of self-efficacy. Thus, to increase females' interest in using specific applications and tools, teachers could integrate these ICT in learning activities, which may help reduce the gender gap in ICT literacy. Future research may need to further investigate adolescents' ICT preferences and experiences with ICT outside school and their role in the acquisition of ICT literacy. In addition, the self-efficacy scales used in ICT literacy-related studies are based on self-reports, and they may not be assessing adolescents' ICT self-efficacy appropriately. Hence, observational studies may be required to appropriately assess students' ICT self-efficacy.

Based on international literature, the chapter discussed the vital role of teachers, schools and education policies to support the use of technologies in the classroom, and help adolescents develop skills needed for the future. Likewise, the UAE government has implemented several school reforms and revamped schools' technological infrastructure to equip Emiratis with the knowledge and skills needed for life and future jobs. A recent OECD report (2020, 2021) suggested encouraging results concerning ICT use in UAE schools and teacher professional development at the secondary level. However, there is limited research on the effectiveness of ICT integration in the classroom and its impact on ICT literacy acquisition. A few studies were conducted at private schools. Tamim (2013), for example, found that while teachers were aware of the benefits of using ICT in the classroom, they lacked an in-depth understanding of pedagogical approaches to integrate these tools. Similar research can be extended to public schools to learn how teachers integrate ICT.

Recommendations for Practice

ICT represents dual goals that support both teaching and learning and the development of ICT literacy (OECD, 2018), which may help increase female participation in STEM-related fields (Chavatzia, 2017). However, ICT literacy must be situated within a more holistic approach that considers other challenges influencing females to choose careers in STEM (Abe & Chikoko, 2020; Almeda & Baker, 2020; Senkbeil et al., 2013). Similarly, in the UAE context, there are challenges, such as students' proficiency with the English language and the capability of STEM teachers influencing Emiratis' choices of STEM subjects or career paths (Makhmasi et al., 2012). The UAE reforms will need to overcome these challenges to improve female Emiratis' interest in STEM fields and decrease the gender gap that persists in the country (Patterson et al., 2021).

Meanwhile, the UAE government's initiatives described in this chapter provide a supporting structure to help adolescents develop the required future skills. With a robust and systematic research agenda, these initiatives and policies can be strengthened and informed by the latest technological developments and pedagogic innovations. In addition, as suggested in the review of the studies, students are acquiring ICT skills outside school due to the informal use of tools and applications at home (Cussó-Calabuig, 2018; Lim & Meier, 2011; Zhong, 2011) which may be the case with Emirati students. As noted by Drabowicz (2014, p. 110),

> computer games and the Internet are an extremely important part of contemporary adolescents' so-called informal learning environment…the fact that girls lag behind boys in the frequency of playing computer games…might contribute in the future to the decrease in number of girls pursuing STEM subjects and entering traditionally male-dominated STEM-related professions…thus slowing the quiet revolution in the position and role of women in society.

Therefore, teachers should incorporate tools and applications used by students outside classes such as games and programming into their teaching to help increase females' interest in ICT, enhance ICT self-efficacy and decrease imbalanced use of technologies. In addition, a community of practice can be a valuable approach to complement other existing types of professional development organised in schools. A community of practice facilitates teacher professional learning and growth and helps them make changes to their practice with the support of community peers (Trust & Horrocks, 2017). It is recommended that school leaders encourage opportunities for all teachers to participate in continuous activities within the community of practice to address diverse needs, sharing helpful practices, experiences and approaches to improve gender gaps in using technologies to develop ICT literacy.

Recommendations for Future Research

Based on the review provided in this chapter, the following research agenda covering the areas listed below are recommended in the UAE context to enhance existing policies, initiatives and teaching practices in secondary schools:

1. Conduct systematic studies in schools using a range of research methodologies to investigate the effectiveness of ICT integration in the classroom, exploring pedagogical approaches, including strategies that help develop adolescents' ICT literacy skills such as searching and evaluating information (Fraillon et al., 2019).
2. Research the implementation of technologies and their association with ICT literacy skills and STEM orientations. For example, of the few studies implemented in the UAE, Ali et al. (2019) provided valuable findings to public schools. They showed significant associations between 3D printing and Emirati female interest in pursuing STEM-related careers.
3. Implement large-scale studies to investigate ICT-related individual characteristics, including ICT self-efficacy, attitudes, interests and ICT usage both inside and outside schools to understand usage patterns and gender inequalities. As noted by Lim and Meier (2011), these studies are needed to identify factors contributing to gender inequalities and enable schools and teachers to create optimal learning environments for all.
4. Create opportunities for school-university partnerships to plan and implement joint research projects to investigate teaching practices and ICT implementation. This type of partnership is not new but has grown in importance in recent years, where school and university professionals work together to improve school practices and policies (Day et al., 2021). In addition, school-university partnerships increase the likelihood of practical research approaches to improve teaching and learning (Bernay et al., 2020).

References

Abe, E. N. & Chikoko, V. (2020). Exploring the factors that influence the career decision of STEM students at a university in South Africa. *International Journal of STEM Education*, *7*(60). https://doi.org/10.1186/s40594-020-00256-x.

ACARA (2012). National assessment program: ICT literacy years 6 and 10. Australian Curriculum, Assessment and Reporting Authority. https://www.nap.edu.au/_resources/NAP_ICTL_2011_Public_Report_Final.pdf.

ACARA. (2014). National assessment program: ICT literacy years 6 and 10. Australian Curriculum, Assessment and Reporting Authority. https://www.nap.edu.au/_resources/D15_8761__NAP-ICT_2014_Public_Report_Final.pdf.

Alaleeli, S., & Alnajjar, A. (2020). The arab digital generation's engagement with technology: The case of high school students in the UAE. *Journal of Technology and Science Education, 10*(1), 159–178. https://doi.org/10.3926/jotse.

Aesaert, K., van Braak, J., Van Nijlen, D., & Vanderlinde, R. (2015). Primary school pupils' ICT competences: Extensive model and scale development. *Computers & Education, 81*, 326–334. https://doi.org/10.1016/j.compedu.2014.10.021.

Ali, N., Areepattamanni, S., Santos, I. M., & Khine, M. (2019). Technology and interest in STEM careers among female elementary students. In P. H. Bull & J. Keengwe (Eds). *Handbook of research on innovative digital practices to engage learners* (138–157). IGI Global. https://doi.org/10.4018/978-1-5225-9438-3.ch007.

Almeda, M. V., & Baker, R. S. (2020). Predicting student participation in STEM careers: The role of affect and engagement during middle School. *Journal of Educational Data Mining, 12*(2), 33–47. https://doi.org/10.5281/zenodo.4008054.

Almekhlafi, A. G., & Almeqdadi, F. A. (2010). Teachers' perceptions of technology integration in the United Arab Emirates school classrooms. *Educational Technology & Society, 13*(1), 165–175.

Atit, K., Power, J.R., Veurink, N., Uttal, D. H., Sorby, S., Panther, G., Msall, C., Logan, F. & Carr, M. (2020). Examining the role of spatial skills and mathematics motivation on middle school mathematics achievement. *International Journal of STEM Education*, 7(38). https://doi.org/10.1186/s40594-020-00234-3.

Bandura, A. (1986). *Social foundations of thought and action: A social cognitive theory*. Prentice Hall.

BECTA. (2008). How do boys and girls differ in their use of ICT? Research Report. https://dera.ioe.ac.uk/8318/1/gender_ict_briefing.pdf.

Bejaković, P., & Mrnjavac, Ž. (2020). The importance of digital literacy on the labour market. *Employee Relations, 42*(4), 921–932. https://doi.org/10.1108/ER-07-2019-0274.

Bernay, R., Stringer, P., Milne, J., & Jhagroo, J. (2020). Three models of effective school–university partnerships. *New Zealand Journal of Educational Studies, 55*(1), 133–148. https://doi.org/10.1007/s40841-020-00171.

Borokhovski, E., Pickup, D., El Saadi, L., Rabah, J., & Tamim, R. M. (2018). *Gender and ICT: Meta-analysis and systematic review*. The Commonwealth of Learning.

Cai, Z., Fan, X., & Du, J. (2017). Gender and attitudes toward technology use: A metaanalysis. *Computers & Education, 105*, 1–13. https://doi.org/10.1016/j.compedu.2016.11.003.

Casey, L., & Bruce, B. C. (2011). The practice profile of inquiry: Connecting digital literacy and pedagogy. *E-Learning and Digital Media, 8*(1), 76–85. https://doi.org/10.2304/elea.2011.8.1.76.

Chavatzia, T. (2017). Cracking the code: girls' and women's education in science, technology, engineering and mathematics (STEM). UNESCO. http://unesdoc.unesco.org/images/0025/002534/253479e.pdf.

Christoph, G., Goldhammer, F., Zylka, J., & Hartig, J. (2015). Adolescents' computer performance: The role of self-concept and motivational aspects. *Computers & Education, 81*, 1–12. https://doi.org/10.1016/j.compedu.2014.09.004.

Claro, M., Preiss, D. D., Martín, S., Jara, I., Hinostroza, H., Valenzuela, S., Cortes, F., & Nussbaum, M. (2012). Assessment of 21st century ICT skills in Chile: Test design and results from high school level students. *Computers and Education, 59*(3), 1042–1053. https://doi.org/10.1016/j.compedu.2012.04.004.

Colley, A., & Comber, C. (2003). Age and gender differences in computer use and attitudes among secondary school students: What has changed? *Educational Research, 45*(2), 155–166. https://doi.org/10.1080/0013188032000103235.

Cussó-Calabuig, R., Farran, X. C., & Bosch-Capblanch, X. (2018). Effects of intensive use of computers in secondary school on gender differences in attitudes towards ICT: A systematic review. *Education and Information Technologies, 23*(5), 2111–2139. https://doi.org/10.1007/s10639-018-9706-6.

Davis, N., Eickelmann, B., & Zaka, P. (2013). Restructuring of educational systems in the digital age from a co-evolutionary perspective. *Journal of Computer Assisted Learning, 29*(5), 438–450. https://doi.org/10.1111/jcal.12032.

Day, C., Gu, Q. Townsend, A., & Holdich, C. (2021). *School-University partnerships in action: The promise of change.* Routledge.

Dlamini, R. & Mbatha, K. (2018). The discourse on ICT teacher professional development needs: The case of a South African teachers' union. *International Journal of Education and Development using Information and Communication Technology, 14*(2), 17–37. https://www.learntechlib.org/p/184684/.

Dogan B., & Robin B. (2015). Technology's role in Stem education and the Stem SOS model. In Sahin, A. (Eds.), *A practice-based model of STEM teaching* (pp. 77–94). Sense Publishers. https://doi.org/10.1007/978-94-6300-019-2_6.

Drabowicz, T. (2014). Gender and digital usage inequality among adolescents: A comparative study of 39 countries. *Computers & Education, 74*, 98–111. https://doi.org/10.1016/j.compedu.2014.01.016.

Eastin, M. S., & LaRose, R. (2006). Internet self-efficacy and the psychology of the digital divide. *Journal of Computer Mediated Communication, 6*(1). https://doi.org/10.1111/j.1083-6101.2000.tb00110.x.

Eickelmann, B. (2011). Supportive and hindering factors to a sustainable implementation of ICT in schools. *Journal for Educational Research Online, 3*(1), 75–103. https://doi.org/10.25656/01:4683.

Ertmer, P. A. (1999). Addressing first- and second-order barriers to change: Strategies for technology integration. *Educational Technology Research and Development, 47*(4), 47–61. https://doi.org/10.1007/BF02299597.

ETS. (2002). *Digital transformation: A framework for ICT literacy.* Educational Testing Service.

Fairlie, R. W. (2015). Do boys and girls use computers differently, and does it contribute to why boys do worse in school than girls? https://www.econstor.eu/bitstream/10419/120955/1/dp9302.pdf.

Ferrari, A. (2013). DIGCOMP: A framework for developing and understanding digital competence in Europe. In Y. Punie & B. N. Brečko (Eds.), *JRC scientific and policy reports.* Publications Office of the European Union.

Fraillon, J., Ainley, J., Schulz, W. Friedman, T., & Duckworth, D. (2019). Preparing for life in a digital world. In *IEA International Computer and Information Literacy Study 2018 International Report*. Springer Open.

Fraillon, J., Ainley, J., Schulz, W., Friedman, T., & Gebhardt, E. (2014). Preparing for life in a digital age. In *The IEA International Computer and Information Literacy Study international report*. Springer. https://www.springer.com/gp/book/9783319142210.

Fraillon, J., Schulz, W., & Ainley, J. (2013). International computer and information literacy study: Assessment framework. https://research.acer.edu.au/ict_literacy/9.

Fraillon, J., Schulz, W., Gebhardt, E., & Ainley, J. (2015). National assessment program: ICT literacy years 6 & 10. Australian Curriculum, Assessment and Reporting Authority (ACARA). https://research.acer.edu.au/ict_literacy/11.

Gebhardt, E., Thomson, S., Ainley, J., & Hillman, K. (2019). *Gender differences in computer and information literacy: An in-depth analysis of data from ICILS*. Springer Open.

Gerick, J., Eickelmann, B., & Bos, W. (2017). School-level predictors for the use of ICT in schools and students' CIL in international comparison. *Large-Scale Assessment Education, 5*(5), 1–13. https://doi.org/10.1186/s40536-017-0037-7.

Gnambs, T. (2021). The development of gender differences in information and communication technology (ICT) literacy in middle adolescence. *Computers in Human Behaviour, 114*. https://doi.org/10.1016/j.chb.2020.106533.

Hatlevik, O. E., Throndsen, I., Loi, M. G. B., & Gudmundsdottir, G. B. (2018). Students' ICT self-efficacy and computer and information literacy: Determinants and relationships. *Computers & Education, 118*, 107–119. https://doi.org/10.1016/j.compedu.2017.11.011.

Hidi, S., & Renninger, K. A. (2006). The four-phase model of interest development. *Educational Psychologist, 41*(2), 111–127. https://doi.org/10.1207/s15326985ep4102_4.

Hohlfeld, T., Ritzhaupt, A., & Barron, A. (2013). Are gender differences in perceived and demonstrated technology literacy significant? It depends on the model. *Educational Technology Research & Development, 61*(4), 639–663. https://doi.org/10.1007/s11423-013-9304-7.

International Literacy Association. (2017). Overcoming the digital divide: Four critical steps. Literacy Leadership Brief. https://www.literacyworldwide.org/docs/default-source/where-we-stand/ila-overcoming-digital-divide-brief.pdf.

International Society for Technology in Education (ISTE). (2016). ISTE Standards. https://cdn.iste.org/www-root/PDF/ISTE%20Standards-One-Sheet_Combined_09-2021_vF3.pdf.

Iroaganachi1, M., Opeke, R., Babalola, Y., & Soyemi, O. (2020). Digital literacy skills and STEM career path choice intentions among junior secondary school girls in North-Central Nigeria. *Journal of Informatics & Communication Technology, 8*(2). https://journals.covenantuniversity.edu.ng/index.php/cjict/article/view/2439.

Jaeggi, S. M., Buschkuehl, M., Shah, P., & Jonides, J. (2014). The role of individual differences in cognitive training and transfer. *Memory & Cognition, 42*(3), 464–480. https://doi.org/10.3758/s13421-013-0364-z.

Jan, S. (2018). Gender, school and class wise differences in level of digital literacy among secondary school students in Pakistan. *Issues and Trends in Learning Technologies, 6*(2). https://doi.org/10.2458/azu_itet_v6i2_jan.

Kozma, R. (2008). Comparative analysis of policies for ICT in education. In J. Voogt & G. Knezek (Eds.), *International handbook of information technology in primary and secondary education* (pp. 1083–1096). Springer.

Kuhlemeier, H., & Hemker, B. (2007). The impact of computer use at home on students' internet skills. *Computers & Education, 49*(2), 460–480. https://doi.org/10.1016/j.compedu.2005.10.004.

Kunina-Habenicht, O., & Goldhammer, F. (2020). ICT Engagement: A new construct and its assessment in PISA 2015. *Large-Scale Assessments in Education, 8*,. https://doi.org/10.1186/s40536-020-00084-z.

Law, N., Pelgrum, W. J., & Plomp, T. (Eds.). (2008). *Pedagogy and ICT use in schools around the world: Findings from the IEA SITES 2006 study*. CERC-Springer.

Lennon, M., Kirsch, I., Von Davier, M., Michael, W., & Kentaro, Y. (2003). *Feasibility study for the PISA ICT literacy assessment*. OECD.

Li, S., Yamaguchi, S., Sukhbaatar, J., & Takada, J. (2016). The Influence of teachers' professional development activities on the factors promoting ICT integration in primary schools in Mongolia. *Education Sciences, 9*, 1–18.

Lim, K. & Meier, E. (2011). Different but similar: Computer use patterns between young Korean males and females. *Education Technology Research and Development, 59*(4), 575–592. https://www.jstor.org/stable/41414959.

Lubinski, D. (2010). Spatial ability and STEM: A sleeping giant for talent identification and development. *Personality and Individual Differences, 49*(4), 344–351. https://doi.org/10.1016/j.paid.2010.03.022.

Makhmasi, S., Zaki, R., Barada, H., & Al-Hammadi, Y. (2012). Students' interest in STEM education: A survey from the UAE. In *Proceedings of the 2012 IEEE Global Engineering Education Conference (EDUCON)* (pp. 1–3). IEEE. https://doi.org/10.1109/EDUCON.2012.6201144.

Ministry of Education Singapore. (2021). Educational technology plan. https://www.moe.gov.sg/education-in-sg/educational-technology-journey/edtech-plan.

Ministry of Education UAE. (2019). The UAE ministry of education showcases its innovative technology as guest of honor at EDUTECH Korea. https://www.moe.gov.ae/En/MediaCenter/News/Pages/Edutech1.aspx.

Ministry of Education UAE. (2021). Quality education. https://u.ae/en/about-the-uae/leaving-no-one-behind/4qualityeducation.

Mullis, I. V. S., Martin, M. O., Foy, P., & Hooper, M. (2017). *ePIRLS 2016 International Results in Online Informational Reading*. http://timssandpirls.bc.edu/pirls2016/international-results/.

OECD. (2011). *PISA 2009 results: Students online*. OECD Publishing. https://doi.org/10.1787/9789264112995-en.

OECD. (2015). Students, computers, and learning: Making the connection. Author.

OECD. (2018). PISA 2021 ICT framework. https://www.oecd.org/pisa/sitedocument/PISA-2021-ICT-Framework.pdf.

OECD. (2019). The role of education and skills in bridging the digital gender divide evidence from APEC economies. https://www.oecd.org/sti/education-and-skills-in-bridging-the-digital-gender-divide-evidence-from-apec.pdf.

OECD (2020). Teaching in the United Arab Emirates: 10 lessons from TALIS. https://www.oecd.org/education/talis/Teaching_in_the_UAE-10_Lessons_from_TALIS.pdf.

OECD. (2021). Teachers getting the best out of their students: From Primary to upper secondary education. *TALIS, OECD Publishing*. https://doi.org/10.1787/5bc5cd4e-en.

Patterson, L., Varadarajan, D. S., & Saji Salim, B. (2021). Women in STEM/SET: Gender gap research review of the United Arab Emirates (UAE): A meta-analysis. *Gender in Management, 36*(8), 881–911. https://doi.org/10.1108/GM-11-2019-0201.

Punter, R. A., Meelissen, M. R., & Glas, C. A. (2017). Gender differences in computer and information literacy: An exploration of the performances of girls and boys in ICILS 2013. *European Educational Research Journal, 16*(6), 762–780. https://doi.org/10.1177/1474904116672468.

Ritzhaupt, A. D., Liu, F., Dawson, K., & Barron, A. E. (2013). From differences in student information and communication technology literacy based on socio-economic status, ethnicity, and gender: Evidence of a digital divide in Florida schools. *Journal of Research on Technology in Education, 45*(4), 291–307. https://doi.org/10.1080/15391523.2013.10782607.

Roblyer, M. D, & Hughes, J. E. (2019). *Integrating educational technology into teaching: Transforming learning across disciplines* (8 Ed). Pearson.

Scherer, R., & Siddiq, F. (2015). Revisiting teachers' computer self-efficacy: A differentiated view on gender differences. *Computers in Human Behavior, 53*, 48–57. https://doi.org/10.1016/j.chb.2015.06.038.

Senkbeil, M. (2021). ICT-related variables as predictors of ICT literacy beyond intelligence and prior achievement. *Education and Information Technologies*. https://doi.org/10.1007/s10639-021-10759.

Senkbeil, M., Ihme, J. M., & Wittwer, J. (2013). The test of technological and information literacy (TILT) in the national educational panel study: Development, empirical testing, and evidence for validity. *Journal for Educational Research Online 5*(2), 139–161. https://doi.org/10.25656/01:8428.

Siddiq, F., & Scherer, R. (2019). Is there a gender gap? A meta-analysis of the gender differences in students' ICT literacy. *Educational Research Review, 27*, 205–217. https://doi.org/10.1016/j.edurev.2019.03.007.

Tamim, R. (2013). Technology integration in the UAE schools: Current status and way forward. In F. Albadri (Ed.), *Information systems applications in the Arab education sector* (pp. 23–38). IGI Global. https://doi.org/10.4018/978-1-4666-1984-5.ch002.

Teo, T. (2008). Assessing the computer attitudes of students: An Asian perspective. *Computers in Human Behavior, 24*(4), 1634–1642. https://doi.org/10.1016/j.chb.2007.06.004.

Tømte, C. (2011). Challenging our views on ICT, gender and education. *Nordic Journal of Digital Literacy, 6*(4), 309–324. https://doi.org/10.18261/ISSN1891-943X-2011-special_issue-04.

Tømte, C., & Hatlevik, O. E. (2011). Gender-differences in self-efficacy ICT related to various ICT-user profiles in Finland and Norway. How do self-efficacy, gender and ICT-user profiles relate to findings from PISA 2006. *Computers & Education, 57*(1), 1416–1424. https://doi.org/10.1016/j.compedu.2010.12.011.

Trust, T., & Horrocks, B. (2017). 'I never feel alone in my classroom': Teacher professional growth within a blended community of practice. *Professional Development in Education, 43*(4), 645–665. https://doi.org/10.1080/19415257.2016.1233507.

Tsai, M. J., & Tsai, C. C. (2010). Junior high school students' Internet usage and self-efficacy: A re-examination of the gender gap. *Computers & Education, 54*(4), 1182–1192. https://doi.org/10.1016/j.compedu.2009.11.004.

UNESCO. (2018). Building tomorrow's digital skills: What conclusions can we draw from international comparative indicators? Working Papers on Education Policy. The United Nations Educational, Scientific and Cultural Organization.

UNESCO. (2019). Exploring STEM competences for the 21st Century. https://unesdoc.unesco.org/ark:/48223/pf0000368485.

UNICEF (2020). Reimagining girls' education through STEM. https://www.unicef.org/media/84046/file/Reimagining-girls-education-through-stem-2020.pdf.

van Deursen, A. J. A. M., & van Diepen, S. (2013). Information and strategic Internet skills of secondary students: A performance test. *Computers & Education, 63*, 218–226. https://doi.org/10.1016/j.compedu.2012.12.007.

van Dijk, J. A. G. M. (2006). Digital divide research, achievements and shortcomings. *Poetics, 34*, 221–235. https://doi.org/10.1016/j.poetic.2006.05.004.

van Laar, E., van Deursen, A. J. A. M., van Dijk, J. A. G. M., & de Haan, J. (2017). The relation between 21st-century skills and digital skills: A systematic literature review. *Computers in Human Behavior, 72*, 577–588. https://doi.org/10.1016/j.chb.2017.03.010.

Vekiri, I. (2008). ICT(s) and socialization: The role of the school and teachers. Paper prepared for the OECD Expert meeting on Gender, ICT and Education. https://www.oecd.org/education/ceri/40832756.pdf.

Wenger, E. (1998). *Communities of practice: Learning, meaning, and identity*. Cambridge University Press.

World Economic Forum. (2016). The future of jobs employment, skills and workforce strategy for the fourth industrial revolution. https://www3.weforum.org/docs/WEF_Future_of_Jobs.pdf.

Zhong, Z. J. (2011). From access to usage: The divide of self-reported digital skills among adolescents. *Computers & Education, 56*, 736–746. https://doi.org/10.1016/j.compedu.2010.10.016.

Zylka, J., Christoph, G., Kröhne, U., Hartig, J., & Goldhammer, F. (2015). Moving beyond cognitive elements of ICT literacy. First evidence on the structure of ICT engagement. *Computers in Human Behavior, 53*, 149–160. https://doi.org/10.1016/j.chb.2015.07.008.

Chapter 7
Understanding of Environmental Issues Across Two Gulf Countries: Do Girls Know More Than Boys in UAE Schools?

Dean Cairns

Abstract The Organization for Economic Cooperation and Development (OECD) is a forum that has a unique influence on governments across the world regarding economic, social and environmental issues. Through the PISA 2015 survey, the OECD measured both students' knowledge and awareness of current environmental challenges. Across the countries involved in the PISA 2015 cycle, environmental awareness was associated with significantly increased scientific literacy scores. In the UAE and Qatar, overall reported environmental awareness in government schools was close to the OECD average. Girls reported significantly higher levels of environmental awareness across both countries than boys. Regression analysis revealed similar positive relationships between environmental awareness and scientific literacy for the UAE and Qatar, equivalent to nearly 9 months of additional schooling ($B = 17.11$, $SE = 1.22$, $p < 0.001$ and $B = 19.00$, $SE = 0.86$, $p < 0.001$, respectively). However, statistically significant differences in the strength of the relationships between girls and boys in both countries were not observed. In this chapter, suggestions for improvements to curriculum, practice and science education (particularly for boys) in government schools are provided to improve the environmental awareness of these students.

Keywords Environmental awareness · Scientific literacy · PISA 2015 · Gender

Introduction

The reported environmental challenges currently facing people across the globe consist of issues such as biodiversity loss, plastic pollution, deforestation, air pollution, climate change and ocean acidification (*11 Biggest Environmental Problems of 2021*, 2021, August 5). This far-reaching list of problems (which is by no means complete) extends to further secondary issues such as sea level rise (associated with ocean warming) or coral bleaching (linked to decreases in ocean pH). By way of

D. Cairns (✉)
Emirates College for Advanced Education, Abu Dhabi, UAE
e-mail: dcairns@ecae.ac.ae

M. Dickson et al. (eds.), *Gender in STEM Education in the Arab Gulf Countries*,
https://doi.org/10.1007/978-981-19-9135-6_7

an example, the Intergovernmental Panel on Climate Change (IPCC) provides periodic systematic reviews of the scientific research literature to inform governments and policymakers regarding the impacts and implications of climate change (*The Intergovernmental Panel on Climate Change*, n.d.). In their most recent report, the physical science working group drew attention to rapid rises in greenhouse gases such as carbon dioxide, methane, and nitrous oxide and unequivocally attributed these rises to human activity (IPCC, 2021). The report also stated that it is likely that humans have caused the global surface temperature to rise by 1.07 °C and are likely responsible for increased precipitation rates over land. Furthermore, human influence is very likely behind glacial retreat, global upper ocean temperature increases, global mean sea level rises, and changes to the land biosphere (increased average growing seasons due to poleward expansion of climate zones).

Possible solutions to each of these environmental concerns lie within the domains not only of science but also of the social sciences and economics. As such, science education has a role to play, particularly in terms of educating the next generation of adults that will also need to face (and hopefully tackle) these challenges. This chapter will focus on the self-reported environmental awareness of 15-year-old students in government schools in the United Arab Emirates (UAE) and Qatar, consisting of awareness of greenhouse gas emissions, genetically modified organisms, nuclear waste, deforestation, air pollution, mass extinctions, and water shortages as identified in the Environmental Awareness scale included in the student questionnaire in the PISA 2015 cycle (OECD, 2017b). These views will be analysed in terms of the relative levels of environmental awareness of these students and the relationships between these views and their performance in the PISA 2015 science assessment. Gender differences will also be examined, as previous research has demonstrated that female students in this region demonstrate higher levels of environmental awareness than males (Al-Naqbi & Alshannag, 2018).

Literature Review

The interconnectedness of the environmental, economic and social domains is encompassed by conceptualizations of education for sustainable development (ESD). The United Nations Educational, Scientific and Cultural Organization (UNESCO) explains that ESD, "empowers learners of all ages with the knowledge, skills, values and attitudes to address the interconnected global challenges we are facing, including climate change, environmental degradation, loss of biodiversity, poverty and inequality" (*Education for Sustainable Development*, n.d.). The implication here is that improved environmental outcomes are dependent on quality education for all. Accordingly, 193 states (https://sustainabledevelopment.un.org/memberstates.html; including the UAE and Qatar) have committed to the United Nation's sustainable development goals (SDGs), which include SDG 4—Quality Education (a commitment to inclusive and equitable education provision). The UAE is currently a member of the UNESCO SDG 4 education 2030 steering committee (United Arab Emirates Ministry of Education, 2018).

The challenge of these educational commitments is to provide educational experiences that develop students' knowledge and awareness of environmental issues in the hope that these then lead to environmentally friendly behaviours and informed actions that follow through into adulthood. Although, as already implied, environmental education is essentially cross-curricular in nature, and science education has an important role in developing both the cognitive (scientific principles underpinning ESD) and the affective (emotions related to nature and the environment) domains. Indeed, Littledyke (2008) presented a model of environmental science education that included strategies for bridging the cognitive and affective domains through teacher modelling of biophilic behaviour, active learning through constructivist pedagogies, engagement with natural environments and living organisms, and curricular activities that demonstrated the complexity of environmental effects of certain human behaviours. However, threats to the effectiveness of these approaches stem from the negative relationships that many students have with learning science in schools, particularly in later grades (Ing & Nylund-Gibson, 2017). Supporting and encouraging early positive dispositions to science education, thus encompassing environmental education, are essential whilst there is a degree of malleability in terms of attitudes to learning in these subjects (Ing & Nylund-Gibson, 2017). Considering these challenges, several countries around the world include explicit teaching related to environmental science in their science curriculums, often as early as in the primary school phase.

Environmental Science in the Curriculum

Environmental science and ESD can be found in various science curricula across the globe. For instance, in the science curriculum for Canada, grades 1–8, the first overarching goal of the curriculum is "to relate science and technology to society and the environment" (Ontario Ministry of Education, 2007, p. 3). Further to this, the curriculum is underpinned by fundamental science concepts, one of which is identified as sustainability and stewardship. Clearly, this curriculum is intended to develop links between scientific concepts and their implications for society from within the science curriculum itself. The current National Curriculum for England does not have such overarching goals relating to sustainability (the word is not included in the documentation at all). However, students are taught to understand the impacts of human activities such as pollution and deforestation, starting in the primary years (Department for Education, 2013). In a report published by the Environmental Education Research group at King's College London, Glackin and King (2018) noted the absence of environmental education in the National Curriculum for England and recommended the development of coherent national policies that recognise the multi-faceted nature of environmental education that ensure students have the opportunities to develop broad understandings of local and global environmental matters.

The education standards for New South Wales in Australia for primary and secondary students include a strand related to attitudes to science whereby students are expected to value the importance and contribution of science in addressing future personal, social and global issues relating to a more sustainable future (Education Standards Authority, n.d.). For example, sustainability in the production of foods and fibres from animals is found in the stage 3 (year 5 and 6) learning outcomes for the Living World strand. The 2021 science syllabuses for Singapore also include elements of environmental science education (Curriculum Planning and Development Division, 2020). The curriculum is organised into five thematic narratives relating to scientific principles and these narratives are supported by (and delivered through) authentic and engaging contexts. This means that contexts such as sustainability and climate change are used as a teaching vehicle for themes such as the diversity of matter or energy transfer.

In the case of the UAE and Qatar, I have used the information provided by each country to the International Association for the Evaluation of Educational Achievement (IEA). This information is required for participation in the Trends in International Mathematics and Science Study (TIMSS). This science and mathematics curriculum-focused assessment is carried out every four years in grades four and eight and, in the 2019 cycle, 64 countries participated. According to the UAE country chapter of the TIMSS 2019 Encyclopedia, in grades 1–4 in the Ministry of Education science curriculum, students are taught about energy transformations and how knowledge of these can support sustainable development (Kelly et al., 2020). In the case of Qatar, grade eight students are taught how environmental change can influence climate and reduce biodiversity (Kelly et al., 2020). It should also be noted that, although environmental science and ESD are clearly represented in these documents, the countries in question also strongly support sustainable development at a policy level beyond the education sector. For example, the UAE Vision 2021 National Agenda focuses on air quality, water supplies, clean energy and green economics (*Sustainable Environment and Infrastructure*, n.d.) and Qatar's National Vision 2030 includes a commitment to meeting all 17 UN SDGs (*Qatar to the Power of Sustainability*, n.d.).

The Effectiveness of Environmental Education

One of the goals of effective environmental science education is that students should develop an increased awareness of environmental issues. This awareness, in turn, should then lead to improved pro-environment behaviours that persist into adulthood. However, although school-based instruction is important to develop an awareness of environmental issues, it may not be sufficient in isolation. For example, a study that used the PISA data for Colombia from the 2009 cycle found only weak associations between the "learning environmental topics" and "environmental awareness" measures (Edsand & Broich, 2020). However, on examination of the learning environmental topics scale, the items only refer to extracurricular and enrichment activities (such as visiting a science center) and are not directly related to the instruction received as part of the formal curriculum. Furthermore, this scale was part of the questionnaire for principals and the environmental awareness scale was administered to students. It is quite possible that the students randomly sampled from the school population had never attended the activities the principal was referring to. For these reasons, weak correlations between the two variables are to be expected and give little information regarding the effectiveness of environmental education, in this case.

The UAE and Qatar

According to Al Blooshi et al. (2020), environmental awareness of UAE residents can is expressed through their concerns regarding energy usage. For example, 56% of respondents to their survey agreed that the *effects* of climate change (warmer climate) had increased their water and energy consumption. Energy and water costs appear to influence environmental awareness, with respondents over 40 providing the strongest agreement relating to questions about climate change, and its effects on their energy consumption. However, this study acknowledges that survey items asking respondents if their energy consumption had increased (in the last 20 years) do not control for confounding variables such as increased access to products such as air conditioning units and increased comfort expectations in modern societies.

Another study analysed the effectiveness of ESD with relation to university-aged students in the UAE (Al-Naqbi & Alshannag, 2018). The findings of this cross-sectional study revealed that the students in the sample exhibited high levels of understanding and had positive attitudes to environmental topics, but these only translated into moderately positive environmentally protective behaviours. It is not clear, however, if the high level of agreement relating to the knowledge domain (for items such as "Human actions are contributing to changes in our atmosphere and climate systems") was developed during their university years or, indeed, if these views were developed during formal education. The authors reported that UAE nationals (who largely tend to enroll in the government school system) exhibited higher levels of knowledge agreement than non-UAE nationals. One posited explanation is that these students developed this knowledge through the more effective delivery of ESD lessons and messages in government schools than in private schools (Al-Naqbi & Alshannag, 2018).

In a study concerning the development of a conceptual framework of environmental science literacy, Khishfe (2014) examined the science curriculum of Qatar to identify aspects of the four domains of environmental literacy (science content knowledge around environmental science, scientific inquiry as a means of inquiring about environmental issues, the nature of science as relating to environmental science and socio-scientific concerns relating to the environment). Three of the four domains were represented in the Qatar science standards, to varying degrees, but the nature of science was almost completely absent. However, there was a strong emphasis on environmental science in the domains of science knowledge and scientific inquiry, with a slightly more marginal coverage of socio-scientific concerns. The study concluded that the Qatar science curriculum is a good starting point for developing environmental science literacy in the Qatari population (Khishfe, 2014).

Environmental Awareness and Science Achievement

In the 2006 and 2015 assessment cycles for PISA the assessment focus (and student questionnaire domain) was science and science education. Included in the student questionnaire was the construct of environmental awareness. This scale was developed for the 2006 cycle to measure the degree of self-reported perceptions of students with regard to key environmental concerns such as greenhouse gases, genetically modified foods, acid rain, nuclear waste and deforestation (OECD, 2009b). The items on the scale were adapted for the 2015 cycle to also include items relating to air pollution, extinctions and water shortages (OECD, 2017b) and the item related to acid rain was removed.

The OECD carries out various analyses of the data captured by the PISA assessments and a report relating to environmental science performance in the 2006 test showed strong, positive relationships between students' reported levels of environmental awareness and their performance in the environmental science questions (OECD, 2009a). However, the relationship for Qatar was not as strong as most other countries, with a one-point increase on the environmental awareness scale relating to only a 10-point increase in their environmental science score (compared with say, 35 points in New Zealand). There is no data for the UAE in this cycle. No such report, providing a country-by-country breakdown of the data, was generated by the OECD for the 2015 cycle but the data relating to environmental awareness and science performance in UAE and Qatari schools is analysed as part of the present study.

In a recent study that used one of the items in the PISA 2015 environmental awareness scale (climate change awareness), Oliver and Adkins (2020) reported that there were strong, positive associations between levels of climate change awareness and science score. However, there were significant variations in the degree of climate change awareness between countries and these variations were not always related to overall PISA performance.

Gender Differences

There is a limited body of research relating to environmental awareness and the moderating effect of gender on, for example, academic performance. In the climate change awareness study discussed previously (Oliver & Adkins, 2020) the authors reported that, across all countries involved in the PISA 2015 assessment cycle (that completed the student questionnaire), females were 4.7% less likely to report being well informed about greenhouse gases in the atmosphere. Interestingly, female students that performed one standard deviation higher in their science score (the study reported standardized regression coefficients) tended to have a lower probability of perceiving themselves as well informed regarding climate change. This is the opposite relationship found for the relationship between students' scores on the full environmental awareness scale and science achievement as reported in the results for the PISA 2006 science assessment (OECD, 2009a). The country specific results for females (regarding climate change awareness and the relationship with science performance) in the UAE and Qatar are not reported in the text of the paper. However, Al-Naqbi and Alshannag (2018) reported that females in the UAE are generally more knowledgeable regarding environmental challenges and concerns than males.

As such, there is a clear gap in the research relating to a systemic analysis of students' awareness of environmental issues in the government-funded school systems of both UAE and Qatar. Furthermore, despite some possible evidence of gender differences in terms of environmental awareness, the available data has not been analysed and reported, particularly in relation to how the reported level of environmental awareness is related to overall science performance in the UAE and Qatar. This is critical information for both countries that have clearly stated aspirations to develop their knowledge-based green economies (through meeting the sustainable development goals), *and* to improve educational outcomes in science for their citizens (particularly in international large-scale assessment like PISA). In order to provide this essential information, the present study will address the following research questions.

Research Questions

1. How do UAE and Qatar government school students' awareness of environmental issues compare with each other and with the international community?
2. How does environmental awareness differ between males and females in UAE and Qatar government schools?
3. What is the relationship between environmental awareness and science achievement in government schools in the UAE and Qatar, and is the relationship the same for males and females?

Methods

Data

The participants in the study were the 15-year-old students that took part in the 2015 PISA cycle attending government-funded schools in Qatar and the UAE. The sample consisted of 5,794 students from schools in Qatar and 4,670 students from schools in the UAE. Overall, the sample consisted of 4155 males (41% of the sample) and 6065 females (59% of the sample). The gender imbalance of the sample is not explained in the PISA documentation for these countries.

Sampling

Sampling was carried out by the OECD to obtain representative samples of the target student population for each country. The target population included 15-year-old students in full-time and part-time education including vocational programmes and students in foreign/international schools (OECD, 2017b). Home schooled students and students studying aboard were not part of the target population. A two-stage stratified sample design was used in each country in this study. The initial stage involved sampling schools systematically from a list of schools eligible to participate in PISA (as described above) with a probability of selection that was proportional to the size of the student population. In other words, larger school populations (with a larger number of 15-year-olds enrolled) had a higher probability of being selected. This method is known as systematic probability proportional to size (PPS) sampling. In addition to this, schools were assigned to specific strata to allow for a more representative sample in terms of the distribution of schools in each country. For example, the strata in the UAE target population schools included schools from each of the 7 emirates, schools offering a range of national and international curriculums, and schools that are publicly and privately owned. In the case of this study, in both countries, only the sampled public schools were included in the analysis.

The second stage of the sampling process involved the selection of students from within the sampled schools. After school selection a list of all eligible students in the school was prepared and (with a target cluster size of 42, for the computer-based test option selected by both countries) where lists of students contained more than the target number, students were selected randomly (with equal probability). Students were not eligible for selection if they met the exclusion criteria. These included students with intellectual disabilities, functional disabilities that may impair their performance, or very poor knowledge of the test language used. Regarding language exclusion, because the countries in this study consist of high numbers of non-Arabic speaking expatriate students, they were provided with an English and Arabic version of the student questionnaire and student assessment (https://www.oecd.org/pisa/test/other-languages/).

These sampling methods mean that, unlike a true random sample, the probability of a student being selected for PISA 2015 varied (by a known amount) depending on how the sampling was carried out for that region. For this reason, survey weights were used to improve the representativeness of each student participant. For example, a school sample design in a country may intentionally oversample a demographic group (e.g., UAE nationals) allowing for accurate parameter estimation in a separate analysis. This over sampling would mean that, without survey weighting, this group's responses would have undue prominence in the dataset and bias the parameter estimates for the target population. As a result, over sampled groups have lower sample weights. In more technical terms, the sampling weight for a given observation for a student is the reciprocal of the probability of including the student in the sample such that the sum of the sampling weights is equal to the population size (Lohr, 2010). Furthermore, the sampling weight at each stage of the stratified sampling method is calculated in the same way (i.e., the probability of the school being selected, *and* the probability of the student being selected). A simplified expression of the weighting calculation used in PISA 2015 is shown in Eq. (7.1) that does not include adjustment factors such as non-participation and unexpectedly large weight values (OECD, 2017b):

$$W_{ij} = w_{2ij} w_{1j} \tag{7.1}$$

where:

W_{ij} is the final student weight for student i in school j;

w_{2ij} is the student (within-school) weight derived from the reciprocal of the probability of selecting student i in school j; and

w_{1j} is the school weight derived from the reciprocal of the probability of selecting the school j.

The school weight was then dependent on the size of the school in comparison to the size of the average school in a stratum.

To ensure the precision of population estimates based on these samples, there was a minimum overall country sample size requirement of 5,250 students (in countries using the computer-based tests). This improved the estimation precision, a value that is inversely related to standard error of estimation and therefore is highly dependent on sample size. This principle also applied to the target cluster size in schools that required a minimum sample size of 20 students. Cluster sizes such as these allowed for reasonably accurate estimates of variance components both within and between schools.

Measures

Predictor Measure

The predictor variable used in this study is a measure of the reported environmental awareness of the respondent. This value is required to address Research Questions 1, 2 and 3 and is drawn from student responses to the context questionnaire completed as part of the PISA 2015 assessment cycle that provided contextual information about the student, their school and the education system of their country. The context questionnaire was developed to provide participating economies with information relating to key policy considerations (OECD, 2017a) such as the human impact on the environment.

The environmental awareness scale used in PISA 2015 required students to rate their knowledge on a four-point scale consisting of a response for low levels of knowledge (I have never heard of this) through to high levels of knowledge (I am familiar with this and would be able to explain it well). Students were asked how informed they were about environmental issues such as greenhouse gases, GMOs, nuclear waste, deforestation, air pollution, mass extinction and water shortages.

The responses to these items (the raw scores) were scaled using item response theory approaches that employed a polytomous model (the generalized partial credit model) allowing for discrimination parameters to vary between items on the scale thus allowing for each item to contribute a different weight to the underlying construct (Muraki, 1992). These item parameters were then used to produce weighted likelihood estimates (WLEs) for each student for each scale (Warm, 1989). The WLEs were then transformed (OECD, 2017b) such that the OECD average was zero and the OECD standard deviation was one, using Eq. (7.2).

$$\theta_j' = \frac{\theta_j - \overline{\theta}_{\text{OECD}}}{\sigma_{\theta(\text{OECD})}} \tag{7.2}$$

where:

θ_j' are the final reported WLE scores;

θ_j are the original WLE scores;

$\overline{\theta}_{\text{OECD}}$ are the mean WLE scores for equally weighted samples from only the OECD countries; and

$\sigma_{\theta(\text{OECD})}$ is the standard deviation of the WLEs for the OECD countries.

The scale reliabilities (Cronbach's alpha values) for the environmental awareness scale were also reported in the PISA documentation as 0.89 and 0.88 for Qatar and the UAE, respectively. These values fall within the accepted ranges for scale reliability (Taber, 2018).

Outcome Measure

The outcome measure used to address Research Question 3, using linear regression modelling, was scientific literacy as measured by the science assessment items in PISA 2015. Scientific literacy, as conceptualised by the OECD, encompasses knowledge about science and technologies that employ scientific knowledge. This is justified by the mechanism of technological advances being driven by scientific advances, and vice versa (OECD, 2017a). This combination of science and technology allows for assessment items, within the framework, that require the respondents to consider the benefits and risks of applying science and technology in their lives and in wider society. In addition to this, scientific literacy is concerned with the processes and practices used to generate scientific knowledge through scientific inquiry.

Thus, scientific literacy is conceptualised as the knowledge of science and technology required by citizens to make informed, critical decisions (rather than only the knowledge required to develop future scientists) and is distilled into three domain specific competencies. These consisted of (1) the ability to explain phenomena scientifically, (2) the ability to evaluate and design scientific inquiry and (3) the ability to interpret data and evidence (OECD, 2017a). Each of these three competencies requires subject knowledge, and the second and third competencies also require knowledge relating to the procedures that produce scientific knowledge (procedural knowledge) and the principles behind the deployment of these practices as methods of knowing and justifying what we believe we know (epistemic knowledge). Understanding of these three types of knowledge (typically developed through school science learning) is required to perform well in the three competencies of scientific literacy.

The content of the assessed science domain spans the major scientific fields of chemistry, biology, physics, and earth and space sciences through item contexts that are authentic, well-known and developmentally appropriate (OECD, 2017a). The procedural knowledge is assessed through items that measure the processes embedded in empirical science inquiry (e.g., variables and controlling variables, repeated measures and presenting and interpreting data). Epistemic knowledge is, in turn, measured by items that question students' knowledge of science-specific constructs (e.g., theories, models, laws, observations, and hypotheses), and the roles of these constructs in justifying scientific knowledge (e.g., how claims about scientific knowledge are supported by data and reasoning, the importance of measurement error, and the purpose of the peer review process).

The administration of the science test items involved the distribution of the selected domain items (after extensive field testing) either on paper, or if via computer with one hour allocated for completing the science section of the test (OECD, 2017b). The complete set of items was not distributed to all students; instead, shorter test forms were used (to reduce the test burden) that consisted of a balanced range of items that covered the above-mentioned content and competency domains with known parameters (item difficulty and discrimination) as determined by item response theory (IRT) methods in the field-testing phase and from items used in previous assessment cycles (OECD, 2017b). The use of short test forms reduces the accuracy of estimations of student performance, but this is deemed acceptable practice in this context, as large-scale assessments are low-stakes assessments for individuals (by design) with no consequences for each individual test taker. To compensate for this lack of accuracy relating to the assessment domain (while still ensuring broad coverage at the population level) other data is taken into consideration when calculating student proficiencies. As such, data from the student background questionnaire (see the subsequent covariates section) are used as part of a latent regression model alongside student estimates based on their responses to items with known item parameters (including the use of covariances in the responses to items from different domains) to estimate student proficiency scores. These scores are reported as a range of values (due to uncertainties relating to the small number of science item responses) known as plausible values. As such, 10 plausible values are calculated for each participant using item calibration based on the IRT model, and population modelling using latent regressions.

Data analysis that used scientific literacy as a variable (such as in this study) thus requires working, not with individual point estimates, but with a distribution of plausible values for each case. Any calculations of point estimates for a given function (such as beta coefficients in regression modeling) must be calculated by running the regression model for each of the plausible values and then obtaining the mean function value. Equally, when calculating the total standard error of such as estimate, the sampling error (arising from obtaining a sample from the population) and measurement error (error induced from method used to calculate proficiencies described above) must be included in the estimate (Rubin, 1987).

The final step entails the transformation of the plausible values to the PISA scale scores. The country means and variances of OECD countries from cycles whereby science was the focus domain were used to calculate transformation coefficients that are then used to adjust the variability of the scale, and the location of the mean on the scale. The nature of this outcome variable, as 10 transformed values drawn from a probability distribution, is accounted for in the data analysis outlined below.

Covariates

The covariates included in the linear regression models used to address Research Question 3 included the index of economic, social and cultural status (ESCS) and gender. The ESCS index is a composite score based on the parental education, parental occupation and home possessions scale. The scores on each of these sub-scales were determined directly from items in the context questionnaire relating to the jobs that their parents have, the level of education or one or more of their parents, and the number of items (such as books) that they have in their homes. This reliabilities of this measure for Qatar and the UAE are low (0.38 and 0.36, respectively).

The variable of gender was also included as this is required to identify gender differences in environmental awareness in order to answer Research Questions 2 and 3. Gender, in this case, is a dichotomous variable where males are coded as zero and females are coded as one.

Results

Statistics and Data Analysis

The data used in this study was drawn from the PISA 2015 database[1] and filtered to include only the data for government-funded schools for the two countries in the study. Weighted descriptive statistics (population estimates) were generated using the International Association for the Evaluation of Educational Achievement's (IEA)

[1] http://www.oecd.org/pisa/data/2015database.

International Database (IDB) Analyzer: Analysis Module (version 4.0.44) (https://www.iea.nl/data) in combination with IBM© SPSS© (version 27) and are provided in Table 7.1. The IEA IDB Analyzer: Analysis Module was developed to provide extra functionality for SPSS (and SAS, although this software is not used in this study) through a graphical user interface that generates syntax for specific SPSS macros that are needed to account for the complex sample designs and multiply imputed test scores found in many large-scale assessment datasets (Sandoval-Hernandez & Carrasco, 2020). In Table 7.1, the scores for scientific literacy are calculated through the combination of the 10 plausible values into a single population estimate, as described previously. As the values in Table 7.1 are estimates based on the sample data, the IEA IDB Analyzer output provides standard errors. However, these values are very small for each estimate and are not included in the table for purposes of clarity. For example, the standard error of estimation for the mean scientific literacy score is reported 0.75. The estimated population mean for environmental awareness is also provided in Table 7.1 and this value was used to address Research Question 1.

To address Research Question 2, the IEA IDB Analyzer: Analysis module was used to calculate the mean environmental awareness of the student population of Qatar and the UAE by gender. The resulting estimates are provided in Table 7.2. This table also includes information relating to missing data for each subgroup, as this is of interest in the analysis later in this chapter.

The final stage of the analysis required the use of the (Pearson) correlation and linear regression functionality of the IEA IDB Analyzer: Analysis Module to answer Research Question 3. The Pearson correlation coefficients for the continuous variables are shown in Table 7.3 and are required to justify the inclusion of these variables in the regression model.

Table 7.1 Weighted descriptive statistics for the measures used by country

Variables	N	Mean	SD
Qatar			
Scientific literacy	7049	386.71	86.10
Environmental awareness[a]	5550	−0.23	1.46
ESCS[a]	6935	0.49	0.85
Gender	7049	0.52	–
UAE			
Scientific literacy	5158	394.41	82.41
Environmental awareness[a]	4670	−0.15	1.34
ESCS[a]	5047	0.33	0.82
Gender	5158	0.54	–

Note Weighted statistics are population estimates that include the adjusted (depending on missing data) final student weight; ESCS = economic, social and cultural status

[a]Reported as a weighted likelihood estimate standardised to a mean of 0 and a SD of 1 for OECD countries

Table 7.2 Weighted descriptive statistics for environmental awareness for each country by gender

Environmental awareness	N	Mean[a]	SD	% missing
Qatar				
Male	2102	−0.30	1.69	36.53
Female	3448	−0.19	1.29	8.05
UAE				
Male	2053	−0.43	1.48	16.37
Female	2617	0.06	1.18	4.28

Note Weighted statistics are population estimates that include the adjusted (depending on missing data) final student weight
[a]Reported as a weighted likelihood estimate standardised to a mean of 0 and a SD of 1 for OECD countries

Table 7.3 Weighted Pearson correlation matrix for the continuous regression variables by country

Variables	a	b	c
Qatar			
a. Scientific literacy	–		
b. Environmental awareness	0.35*	–	
c. ESCS	0.19*	0.18*	–
UAE			
a. Scientific literacy	–		
b. Environmental awareness	0.32*	–	
c. ESCS	0.13*	0.19*	–

Note ESCS = economic, social and cultural status
*$p < 0.01$

The first linear regression model (Model 1) included the predictor variables of environmental awareness, ESCS and gender and the outcome variable of scientific literacy. The purpose of this model was to estimate the relationship between environmental awareness and scientific literacy while accounting for the covariance of ESCS and gender on this outcome variable. In this way it is possible to partial out the unique contribution of varying levels of environmental awareness on the relationship with scientific literacy. The estimates produced by this model are reported in Table 7.4.

Following this, the second linear regression model (Model 2) also included an interaction term between gender and environmental awareness (gender*environmental awareness). The purpose of this model was to test if any identified relationships between environmental awareness and scientific literacy, discovered in Model 1, varied if the respondent was male or female. In other words, is the relationship stronger or weaker for males than it is for females? The results of this model are also shown in Table 7.4.

Table 7.4 Environmental awareness predicting scientific literacy by country

Variables	Model 1		Model 2	
	B	*SE*	*B*	*SE*
Qatar				
Intercept	382.43*	2.03	382.26*	2.06
Environmental awareness	19.00*	0.86	18.38*	1.09
ESCS	13.97*	1.47	13.94*	1.47
Gender	23.20*	2.40	23.51*	2.46
Environmental awareness*Gender	–		1.30	1.57
R^2	0.16		0.16	
UAE				
Intercept	380.20*	4.68	379.29*	4.83
Environmental awareness	17.11*	1.22	14.72*	1.84
ESCS	8.74*	2.09	8.44*	2.07
Gender	33.77*	5.20	34.62*	5.26
Environmental awareness*Gender	–		5.26	2.56
R^2	0.15		0.15	

Note ESCS = economic, social and cultural status
*$p < 0.001$

Discussion

To address Research Question 1 the weighted average reported environmental awareness for government schools in the UAE and Qatar was estimated. As seen in Table 7.1, the mean environmental awareness for students in government schools in Qatar was slightly lower than the OECD average of zero (−0.23). As these scales are standardised to a mean of zero and a standard deviation of one, then this score is 0.23 standard deviations below the OECD average. The mean environmental awareness score for students in government schools in the UAE was also slightly below the OECD average (−0.15). It should first be noted that the OECD average applies to all schools in the member countries and not a subgroup of the schools as reported in this study. For this reason, a direct comparison between the government-funded school systems across the two countries included in this study is more meaningful and, there are only marginal differences observed (−0.23 to −0.15 for Qatar and the UAE, respectively). These values suggest that, on average, the environmental awareness of students in these countries is comparable with OECD countries and indicate that the school curriculum and messaging in wider society is largely successful in developing an OECD equivalent level of environmental awareness in 15-year-old students. However, although it is highly likely that these students' science education is a critical factor, it is not possible to attribute this relatively good knowledge of environmental challenges directly to the teaching and learning taking place in schools.

Certainly, in addition to the school curriculum, through their lifetime, students in these countries are exposed to consistent messaging related to environmental issues in the popular media, in day-to-day life, and from highly respected political leaders. For example, Sheikh Zayed bin Sultan Al Nahyan, the founding father of the UAE, made many declarations supporting the principles that underpin the concept of environmental sustainability, such as in the following quote as part of a speech delivered on the occasion of the 1998 Annual Environment Day:

> We cherish our environment because it is an integral part of our country, our history and our heritage. On land and in the sea, our forefathers lived and survived in this environment. They were able to do so only because they recognized the need to conserve it, to take from it only what they needed to live, and to preserve it for succeeding generation. (Azar & Raouf, 2017, p. 215)

Research Question 2 requires a comparison between the reported environmental awareness of students in governmental schools in these two countries by gender. The results of this analysis are presented in Table 7.2. There is a noticeable difference observed, for both countries, in the levels of reported environmental awareness in males and females. In the case of Qatar, although the overall mean is −0.23 the value for males is reported as −0.30 (and −0.19 for females). This suggests that males in Qatar government schools have on average 0.11 standard deviations lower environmental awareness than females.

Part of the explanation of this disparity may be related to missing data. As shown in Table 7.2, in the case of males in Qatar, 37% of the participants (as a percentage of the total sample size) did not respond to the environmental awareness scale. An analysis of the missing data on a school-by-school basis revealed no pattern in the data by school or region. However, a statistical analysis of the missing values to test for the dependency of missingness on the other variables in the dataset (in this case ESCS and scientific literacy, scales for which there was no missing data) revealed that the missing data was not missing completely at random (MCAR) and was dependent on these variables. This was evidenced from a significant Little's MCAR test in SPSS using the expectation maximization (EM) function in the Missing Values Analysis module, χ^2 (8, $N = 3300$) $= 235.60$, $p < 0.001$. This is a test to see if the missing data patterns for the variables included in the test differ significantly. The null hypothesis is that there are no significant differences between the means of the missing data patterns. Rejection of the null hypothesis indicates that the data is not MCAR (Li, 2013). As such, the implication of this missing data pattern is that students with lower ESCS values (socio-economic status) and lower level of scientific literacy (lower scores on the assessment) were less likely to respond to the environmental awareness scale. This leads to bias in the mean estimates for both Research Question 1 and Research Question 2 for this subgroup (males in government schools in Qatar), resulting in an overestimation of the level of environmental awareness, suggesting that the disparity between males and females is slightly larger than originally reported. Indeed the (unweighted) EM mean scale score for males is estimated to be −0.39 (compared to the raw mean of −0.29) indicating an overestimation of 0.10 standard deviations. In the case of females (with a lower missing data rate) the disparity due to missing data is estimated to be only 0.02 standard deviations.

In the case of the UAE, there is a greater difference between males and females noted in Table 7.2. The weighted mean scale score for environmental awareness is −0.43 for males (almost half of one standard deviation lower than the OECD average of zero) and 0.06 for females, which it is very slightly *above* the OECD average. Again, this data has a high number of missing values in the male sample. In total, 16% of the males in the UAE government school sample did not complete the environmental awareness scale. Again, this data was found to not be MCAR and was dependent on ECSE and scientific literary scores as per the results of Little's MCAR test, χ^2 (5, $N = 2440$) = 72.61, $p < 0.001$. However, the estimated EM means, when accounting for this missing data, were very similar to the unweighted raw means (−0.49 and −0.47, respectively). The missing data for females (only 3.7%), in the case of the UAE, had almost no impact on the estimated EM mean suggesting that the estimated values in Table 7.2, for the UAE data, are unbiased by the missing responses.

As such, there is strong evidence, in both countries, that 15-year-old males studying in government schools report a considerably lower level of environmental awareness than females. In the UAE, the difference in the performance of males and females in the government school sector is relatively well reported. In an early study, Ridge (2010) analysed the quality of teachers and teaching in the government education system and highlighted a marked difference in quality between secondary school experiences by gender. The primary areas of concern in boy's schools are related to overall teacher quality. Teacher quality is explained as being dependent on the characteristics of the teachers and their personal attributes (quality of teacher education, teacher verbal and intellectual aptitude, teacher nationality in relation to the students, and their professional status), school quality (leadership, availability of resources, and general working conditions) and the overall quality of the system the teachers work within. In the majority of these areas teacher quality is likely better in girls' schools than boys' schools and continues to result in a reverse gender gap in both the UAE and Qatar (Marquez et al., 2020, October 27).

In the case of Qatar, similar challenges around teacher quality have been reported in the past (Zellman, 2009). Gender-segregated secondary schools resulted in Qatari boys being taught by Arabic speaking expatriates (non-nationals) who reportedly often lacked the authority required to effectively discipline the students that were exacerbated by job security fears, due to the requirement for yearly contract renewals. Losing their jobs, in this context, resulted in having to leave the country (Zellman, 2009). However, as of 2011, all government schools in Qatar had converted to a charter-school model whereby government-funded schools were designated as independent schools (Said, 2016). These schools had a higher degree of autonomy regarding the allocation of resources and spending, compared with the previous Ministry of Education school model, allowing for more freedom in hiring and promotion practices (Nasser, 2017). However, these reforms resulted in many Qatari teachers leaving the profession due to the diversification of the workforce (in part due to the addition of the requirement to teach certain subjects in English) and increased training requirements leading, again, to likely teacher quality challenges relating to expatriate teachers. The performance of Qatari students in international large-scale assessments such as PISA 2018 remains below the OECD average for subjects such as science, and girls continue to outperform boys (OECD, 2019) by the equivalent of over 1 year of additional schooling (39 points).

This performance gap (also applicable to male and female performance in scientific literacy in the last science cycle in PISA 2015) is likely pertinent to variables such as student environmental awareness, as the item contexts in the PISA 2015 scientific literacy assessment (personal, local/national, global) are applied across questions relating to environmental quality (including biodiversity, sustainability, and pollution), hazards (including climate change impacts) and natural resources (including renewable resources and resource consumption) (OECD, 2017a). This relationship was tested explicitly using linear regression methods to address Research Question 3.

The relationship between environmental awareness and scientific literacy, while accounting for the covariates of ESCS and gender, was tested using linear regression modelling as described in the data analysis section of this study. The results of this analysis are reported in Table 7.4 as Model 1. In the case of Qatar, student-reported levels of environmental awareness were significantly, positively related to scientific literacy ($B = 19.00$, $SE = 0.86$, $p < 0.001$). This suggests that a 1-point increase in environmental awareness is related to a 19-point increase in scientific literacy. According to the OECD one additional year of school is equivalent to 30-points on the PISA scale (Jerrim & Shure, 2016), As such, this 19-point difference is approximately equivalent to 9 months of additional schooling. The results of this analysis are similar for the UAE. Student-reported levels of environmental awareness were again significantly, positively related to scientific literacy ($B = 17.11$, $SE = 1.22$, $p < 0.001$). In this case, a 1-point increase in environmental awareness is related to a 17-point increase in scientific literacy or, more usefully, nearly 7 months of additional schooling. These relationships indicate that students' environmental awareness is an important factor in increasing scientific literacy and, of course, an important factor in developing environmentally informed and responsible citizens.

To address the second part of Research Question 3, an interaction term was introduced to the regression model (Model 2, Table 7.4) to test if the observed relationship between environmental awareness and scientific literacy varies for males and females in both the UAE and Qatar. As can be seen in Table 7.4, in both cases, the interaction effect is not significant (and there is no improvement in the coefficient of determination, R^2). That is, the proportion of variance in the outcome variable that is explained by the predictor variables in Model 2 is not improved over Model 1. As such the relationship is observed as being the same for males and females in both countries.

Conclusions and Implications

In general terms both the UAE and Qatar appear to be successful in developing environmental awareness of adolescents in their respective government school systems.

Students in these schools exhibit levels of environmental awareness that are comparable to the average levels reported across OECD countries. However, when analysing these results by gender, the overall mean environmental awareness is considerably reduced by the male sample. Statistical biases relating to missing data aside, there is a clear disparity between males and females. Although it is not possible to directly attribute this difference to their schooling experience, it is a reasonable assumption that the teaching and learning taking place (particularly in science lessons) is likely to have a direct impact of students' knowledge of socio-scientific concerns around the environmental and human impacts on ecosystems. Furthermore, there is a strong relationship between students' knowledge of these issues and their scientific literacy as measured by PISA 2015.

As such, improving the environmental awareness of boys in the government-funded school sector is a priority with regards to policy and practice. Improving the environmental education (and general education) experience for boys would entail improving the science teaching quality in boys' secondary schools. The teachers responsible for teaching boys in the UAE (for example) are predominantly expatriates (90%; Kippels & Ridge, 2019) usually drawn from neighbouring Arabic speaking nations (to make up for the shortfall in Emirati male teachers), yet the majority of boys studying in these schools are Emirati nationals (81%; as reported in 2014–15). Aside from policies relating to encouraging more Emirati males into the teaching profession, as outlined by Kippels and Ridge (2019) other policy directions should be related to improved recruitment and training of expatriate teachers. For example, although there are high quality programmes available locally and moves are being made to develop the pedagogical content knowledge of teachers, there is currently no teaching qualification requirement for secondary teachers in the government school sector in the UAE. An extensive teacher licencing system has been introduced in Qatar (Nasser, 2017), but the importance of academic qualifications in teaching and learning should not be overlooked. High quality teacher preparation (and training) programmes (at the graduate level) are required that provide effective links between institution-based courses and school-based learning tasks, closely supervised application of pedagogies in schools that make explicit connections between theory and practice, and strong relationships with partner schools that allow (in-service or pre-service) learners to develop effective teaching practices (Darling-Hammond, 2006).

Since a major limitation of this study is the high frequency of missing data, a logical first step would be the collection of data relating to environmental awareness from a representative sample of (in particular) Qatari boys in government-funded schools to provide unbiased estimates for the population under study. Further research in this area could also include a detailed document analysis for each country that would provide recommendations for curriculum development and implementation. Due to the limitations of the data used in the current study (self-reported, cross-sectional, and not longitudinal) other methods of determining students' understanding of environmental concerns should also be employed that include student, teacher and parent interviews, case studies of practice, locally distributed surveys (of students, teachers, and parents), and classroom observations and interventions relating to improved environmental science education.

It is becoming increasingly apparent that a scientifically and environmentally informed populace is important for a nation's economy and for the well-being of its current and future citizens. The measures recommended above would increase understanding of this educational concern and likely lead to improved outcomes in science and improved environmental awareness for students in government schools in the UAE and Qatar, particularly for boys.

References

Biggest Environmental Problems of 2021. (2021, August 5). Retrieved November 7, 2021, from https://earth.org/the-biggest-environmental-problems-of-our-lifetime/.

Al-Naqbi, A. K., & Alshannag, Q. (2018). The status of education for sustainable development and sustainability knowledge, attitudes, and behaviors of UAE University students. *International Journal of Sustainability in Higher Education*. https://doi.org/10.1108/IJSHE-06-2017-0091.

Al Blooshi, L. S., Ksiksi, T. S., Gargoum, A. S., & Aboelenein, M. (2020). Climate change and environmental awareness: A study of energy consumption among the residents of Abu Dhabi, UAE. *Perspectives on Global Development and Technology, 18*(5–6), 564–582. https://doi.org/10.1163/15691497-12341533.

Azar, E., & Raouf, M. A. (2017). *Sustainability in the Gulf: Challenges and opportunities*. Routledge.

Curriculum Planning and Development Division. (2020). *Science syllabuses: Lower secondary express course normal (academic) course*. https://www.moe.gov.sg/-/media/files/secondary/syllabuses/science/2021-science-syllabus-lower-secondary.pdf.

Darling-Hammond, L. (2006). Constructing 21st-century teacher education. *Journal of Teacher Education, 57*(3), 300–314. https://doi.org/10.1177/0022487105285962.

Department for Education. (2013). *National curriculum in England: Science programmes of study*. https://www.gov.uk/government/publications/national-curriculum-in-england-science-programmes-of-study.

Edsand, H.-E., & Broich, T. (2020). The impact of environmental education on environmental and renewable energy technology awareness: Empirical evidence from Colombia. *International Journal of Science and Mathematics Education, 18*(4), 611–634. https://doi.org/10.1007/s10763-019-09988-x.

Education for Sustainable Development. (n.d.). Retrieved November 7, 2021, from https://en.unesco.org/themes/education-sustainable-development.

Education Standards Authority. (n.d.). *Science in kindergarten to year 10*. https://educationstandards.nsw.edu.au/wps/portal/nesa/k-10/learning-areas/science.

Glackin, M. A., & King, H. (2018). Understanding environmental education in secondary schools in England: Report 1: Perspectives from policy. In *King's College London*. King's College London.

Ing, M., & Nylund-Gibson, K. (2017). The importance of early attitudes toward mathematics and science. *Teachers College Record, 119*(5), 1–32. https://doi.org/10.1177/016146811711900507.

The Intergovernmental Panel on Climate Change. (n.d.). Retrieved November 7, 2021, from https://www.ipcc.ch.

IPCC. (2021). Summary for policymakers. In V. Masson-Delmotte, P. Zhai, A. Pirani, S. L. Connors, C. Péan, S. Berger, Y. N. Caud, Chen, L. Goldfarb, M. I. Gomis, M. Huang, K. Leitzell, E. Lonnoy, J. B. R. Matthews, T. K. Maycock, T. Waterfield, O. Yelekçi, R. Yu, & B. Zhou (Eds.), *Climate change 2021: The physical science basis. Contribution of working group I to the sixth assessment report of the intergovernmental panel on climate change*. Cambridge University Press (in press). https://www.ipcc.ch/report/ar6/wg1/downloads/report/IPCC_AR6_WGI_SPM_final.pdf.

Jerrim, J., & Shure, N. (2016). Achievement of 15-year-olds in Northern Ireland: PISA 2015 national report. *London, England: UCL*. https://www.education-ni.gov.uk/sites/default/files/publications/education/ACHIEVEMENT%20OF%2015%20YEAR-OLDS%20IN%20NORTHERN%20IRELAND%20PISA%202015%20NATIONAL%20REPORT..PDF.

Kelly, D. L., Centurino, V., Martin, M. O., & Mullis, I. V. S. (2020). *TIMSS 2019 encyclopedia: Education policy and curriculum in mathematics and science*. https://timssandpirls.bc.edu/timss2019/encyclopedia/index.html.

Khishfe, R. (2014). A reconstructed vision of environmental science literacy: The case of Qatar. *International Journal of Science Education, 36*(18), 3067–3100. https://doi.org/10.1080/09500693.2014.951980.

Kippels, S., & Ridge, N. (2019). The growth and transformation of K–12 education in the UAE. In *Education in the United Arab Emirates* (pp. 37–55). Springer. https://doi.org/10.1007/978-981-13-7736-5_3.

Li, C. (2013). Little's test of missing completely at random. *The Stata Journal, 13*(4), 795–809. https://doi.org/10.1177/1536867X1301300407.

Littledyke, M. (2008). Science education for environmental awareness: Approaches to integrating cognitive and affective domains. *Environmental Education Research, 14*(1), 1–17. https://doi.org/10.1080/13504620701843301.

Lohr, S. L. (2010). *Sampling: Design and analysis (advanced series)*. Brooks/Cole Cengage Learning.

Marquez, J., Lambert, L., Ridge, N., & Walker, S. (2020, October 27). The PISA performance gap between national and expatriate students in the United Arab Emirates. https://doi.org/10.31235/osf.io/rf8hj.

Muraki, E. (1992). A generalized partial credit model: Application of an EM algorithm. *Applied Psychological Measurement, 16*(2), 159–176. https://doi.org/10.1177/014662169201600206.

Nasser, R. (2017). Qatar's educational reform past and future: Challenges in teacher development. *Open Review of Educational Research, 4*(1), 1–19. https://doi.org/10.1080/23265507.2016.1266693.

OECD. (2009a). *Green at fifteen? How 15-year-olds perform in environmental science and geoscience in PISA 2006*. https://www.oecd.org/pisa/pisaproducts/42467312.pdf.

OECD. (2009b). *PISA 2006 technical report*. https://doi.org/10.1787/9789264048096-en.

OECD. (2017a). *PISA 2015 assessment and analytical framework*. https://doi.org/10.1787/9789264281820-en.

OECD. (2017b). *PISA 2015 technical report*. http://www.oecd.org/pisa/data/2015-technical-report/.

OECD. (2019). *PISA 2018 results (volume I)*. https://doi.org/10.1787/5f07c754-en.

Oliver, M. C., & Adkins, M. J. (2020). "Hot-headed" students? Scientific literacy, perceptions and awareness of climate change in 15-year olds across 54 countries. *Energy Research & Social Science, 70*, 101641. https://doi.org/10.1016/j.erss.2020.101641.

Ontario Ministry of Education. (2007). *The Ontario curriculum grades 1–8*. http://www.edu.gov.on.ca/eng/curriculum/elementary/scientec18currb.pdf.

Qatar to the Power of Sustainability. (n.d.). Retrieved November 7, 2021, from https://sustainableqatar.com.

Ridge, N. (2010). *Teacher quality, gender and nationality in the United Arab Emirates: A crisis for boys*. Dubai: Dubai School of Government.

Rubin, D. B. (1987). *Multiple imputation for nonresponse in surveys*. Wiley.

Said, Z. (2016). Science education reform in Qatar: Progress and challenges. *EURASIA Journal of Mathematics, Science and Technology Education, 12*(8), 2253–2265. https://doi.org/10.12973/eurasia.2016.1301a.

Sandoval-Hernandez, A., & Carrasco, D. (2020). Analysing PIAAC data with the IDB analyzer (SPSS and SAS). In *Large-scale cognitive assessment* (pp. 117–148). Springer. https://doi.org/10.1007/978-3-030-47515-4_6.

Sustainable Environment and Infrastructure. (n.d.). Retrieved November 7, 2021, from https://www.vision2021.ae/en/national-agenda-2021/list/environment-circle.

Taber, K. S. (2018). The use of Cronbach's alpha when developing and reporting research instruments in science education. *Research in Science Education, 48*(6), 1273–1296. https://doi.org/10.1007/s11165-016-9602-2.

United Arab Emirates Ministry of Education. (2018, March 20). *UAE gains membership to the SDG4-Education 2030 Steering Committee*. https://www.moe.gov.ae/En/MediaCenter/News/Pages/UAESDG.aspx.

Warm, T. A. (1989). Weighted likelihood estimation of ability in item response theory. *Psychometrika, 54*(3), 427–450. https://doi.org/10.1007/BF02294627.

Zellman, G. L. (2009). *Implementation of the K-12 education reform in Qatar's schools. Monograph*. ERIC. https://www.rand.org/content/dam/rand/pubs/monographs/2009/RAND_MG880.pdf.

Part III
Gendered Representation and Experiences in STEM Education

The six countries of the GCC have made considerable progress, to varying degrees, in at least some of the Global Gender Gap indices related to gender representation in education and labour. The representation of females in all of the GCC countries in primary education has reached parity, or close to it, while Kuwait, Bahrain and the UAE achieved secondary education gender parity. In tertiary education, female enrolment significantly outnumbers male enrolment in all GCC countries, with the gender disparity in Qatar being particularly striking. These figures are not particularly reflected in figures of STEM graduates however, with the exception of Oman, where 55.73% of STEM graduates are female compared to 44.27% male. Other GCC countries still have gender disparities to varying degrees. Of these, Saudi Arabia has the widest gender disparity in STEM graduates (just 36.83% females to 63.18% males). Table 1 presents samples of various indices reproduced from the Global Gender Gap Report (2022). The literacy rate scores are notably a major success story for the GCC countries.

Women in the GCC are among the most highly educated across the Arab World (GGG Report, 2022), yet still gender disparities in the labour market persist. As Table 1 shows, the most noticeable is the gender disparity in the labour force participation rate, and consequently their (lack of) economic participation. Saudi Arabia also has the largest disparity in participation in the labour force and has a long way to go to close its gender gap in multiple indices as a result. Globally gender pay gaps are still large, though the UAE has made impressive attempts in recent years to close this gap, and as 68th out of 146 countries is seen to be far ahead of its GCC neighbours. These all link into the United Nations Sustainable Development Goals (SDGs) ratified in 2016, including SDG 5 ("achieve gender equality and empower all women and girls"), a SDG closely linked to labour participation and workplace rights. The GCC lags behind other countries both globally and in the MENA region with respect to this goal, even though they have raised their index in various other SDGs (such as economic growth and industry).

The UAE tends to outperform the rest of the GCC in many indices, with a top ranking among Arab countries, and a global high ranking in indicators such as the number of women in parliament. In terms of political participation, there are large

Table 1 Global gender gap index indicators (information taken from the World Economic Forum Global Gender Gap Insight Report, 2022)

GCC country	Global gender gap index (ranked out of 146 countries)	Literacy rate score	Labour force participation rate (%)		Enrolment in tertiary education		STEM graduates (%)	
			Female	Male	Female	Male	Female	Male
Bahrain	131	1.0	42.36	83.53	78.51	45.26	41.21	58.79%
Kuwait	130	0.982	47.43	83.30	81.78	43.45	n.a.	n.a.
Oman	139	0.956	28.72	85.04	57.38	35.57	55.73	44.27
Qatar	137	1.0	57.22	95.46	58.77	8.04	n.a.	n.a.
Saudi Arabia	127	0.974	30.94	80.15	73.56	67.93	36.83	63.18
United Arab Emirates	68	0.982	46.54	88	66.35	46.36	41.49	58.51

https://www3.weforum.org/docs/WEF_GGGR_2022.pdf

differences across the GCC countries, with the UAE achieving close too parliamentary level gender parity (0.402), while its bordering neighbour Saudi Arabia remains with a score of 0.0. In 2022, the UAE, Oman and Kuwait had all increased the number of women employed in technical roles, but according to the 2022 Gender Pay Gap Report, only Oman of these three countries has increased the percentage of women holding senior roles.

In the previous section of the book, the ways in which students' experiences of learning STEM subjects within both formal schooling environments and informal learning situations, can impact on students' attitudes, beliefs, etc., were explored. The next section links this now to gendered representation in STEM education. Inequitable representation of females in professorial roles in universities is well documented internationally, explained through the use of interesting metaphors such as the 'leaky pipe', the 'glass ceiling' etc., but this inequity is often intensified in STEM fields. However, relatively scarce literature exists on the actual educational experiences of women who do manage to successfully mount the barriers along this metaphorical pipeline, yet their experiences could provide vital insights into what has been effective, and conversely what has hampered their journeys through formalised schooling systems. The research presented in this section for one of the chapters poses this question: for those minority of women who have reached female senior leadership roles in STEM fields, how have they done this?

Potentially, learning from this kind of empirical research will help to illuminate the types of practices and policies that have supported women to reach such positions. In the same vein, the underrepresentation of females in senior leadership positions is the near-normalised status quo in educational institutions all over the world, yet most empirical research on the subject focuses on regional areas such as Europe, North

America and Australasia. Do the same patterns exist in the Gulf region, and do these patterns in fact appear to be exacerbated in STEM fields specifically? Who are the faculty in STEM fields, and what types of positions and academic ranks do they most commonly habituate? There is much to be learned from listening to narratives of women who have experienced the journey personally as the 'expert witnesses', and much to be learned from gaining a better understanding of the actual make-up of STEM faculty and leadership in a Gulf country, in order to be able to make informed recommendations for trying to overturn gender imbalance in these areas.

Chapter 8
Female STEM Leadership in the Gulf: Journeys Through Education

Martina Dickson and Masada Al Harthi

Abstract The continuing emphasis on the development of a Knowledge Economy in the Gulf Cooperative Council (GCC) intends to transform countries away from being predominantly fossil-fuel dependent. This means that citizens' uptake of science, technology, engineering and mathematics (STEM) degrees and careers, and having highly competent leaders in STEM fields, is of paramount importance. Knowledge and understanding of how to avoid worker attrition in STEM fields is also critical. In the current era of rapid educational transformation in the Gulf, women have an important part to play in the development of their country's educational system. Many young Gulf women are the first of their generation and families to enter into higher education. This chapter charts the narrative journeys of fifteen female leaders who have navigated their way to success in STEM fields, and focuses on the role that schooling and university experiences have played in their lives as they have forged this path. In school, key themes that emerge are the positive impact which a teacher's care and concern about a student has, as well as the capacity of exposure to STEM experiences to ignite an interest and engagement in learning both within and outside of the classroom. All but one of the fifteen leaders interviewed undertook doctoral studies (and most, Masters degrees too) overseas in international universities. This is conceptualised by the women as a cathartic experience of highs and lows, but ultimately one that greatly developed their sense of independence, resilience, self-efficacy and the broadening of their world view as a result of the diversification of experiences outside of their native countries.

Keywords Female leadership · STEM · Success factors · Schooling · University

M. Dickson (✉) · M. Al Harthi
Emirates College for Advanced Education, Abu Dhabi, UAE
e-mail: martina_dickson@hotmail.com

M. Dickson et al. (eds.), *Gender in STEM Education in the Arab Gulf Countries*,
https://doi.org/10.1007/978-981-19-9135-6_8

Introduction

We open this chapter with a brief overview of the current status of females in science, mathematics, engineering and mathematics (STEM) positions globally, including the challenges they face in scaling the echelons of these fields, by way of setting the scene for the rationale for such a study. We then explain the methodology of the current study, present findings of the participants' experiences and how this has impacted their career pathways, and then consider the implications of these to schooling establishments. Much literature on female leadership is dominated by the examination of challenges and barriers which women doubtless encounter on their paths through what remains a male-dominated field, but it is also important to emphasise what has *enabled* their success. These stories can serve as powerful references for all involved in the education of girls and women such as teachers and administrators in schools and universities, but also to women in early STEM careers.

Females in STEM Leadership Positions Globally

The underrepresentation of women in science, technology, engineering and mathematics (STEM) disciplines, which is considered a global phenomenon, is documented extensively in the literature from various perspectives (e.g. Fatourou et al., 2019; Makarem & Wang, 2020; Schmitt & Wilkesmann, 2020; Schmitt et al., 2021). Women have accomplished significant gains in the professional world. However, in STEM arenas, they still represent a minority group with low participation in those male-dominated fields (Goris, 2020; Makarem & Wang, 2020; Schmitt et al., 2021). For instance, in the United States, although women account for half of the workforce, they hold less than 30% of STEM positions, with only 12% in engineering and 26% in computer science (Amon, 2017; Goris, 2020). A similar trend exists in Australia as women represent less than 13% of engineers in the country, while in Germany, the proportion of women in engineering and computer science was 18 and 16% respectively (Fatourou et al., 2019; Schmitt et al., 2021). Women are underrepresented in STEM sectors and leadership positions, and the higher the rank in STEM, the less likelihood of assignment of the position to a woman (Amon, 2017; Dutta, 2018). For example, Botella et al. (2019) reported that women in Europe hold just 21% of high executive positions in technology and 13% in engineering. Such underrepresentation for women in STEM is a multi-faceted issue with different contributing factors influencing women's advancement in their professions. We now look at these factors and how they might hinder or support women's journeys in STEM globally and locally.

Challenges Encountered by Women Leaders in STEM

A significant body of previous studies (Amon, 2017; Babalola et al., 2021; Botella et al., 2019; Dutta, 2018; Goris, 2020; Makarem & Wang, 2020; Minnotte & Pedersen, 2021; Schmitt et al., 2021; Watkins & Mensah, 2019; Wilkins-Yel et al., 2021) has investigated hurdles women face in joining or progressing in STEM occupations. For example, Makarem and Wang (2020) analysed 28 empirical studies and found four common barriers faced by women working in STEM: (a) gendered organisational culture, (b) gender role stereotyping, (c) work-life balance conflict and (d) lack of mentors. These obstacles also appeared in other studies of women working in STEM in both industrial and academic sectors. Women in STEM face negative stereotypes regarding their science abilities and are viewed as less competent than men, impacting peer relationships and promotional opportunities (Goris, 2020). Such gender-stereotyped prejudices affect women's self-image, and some studies reported that even women who hold technical degrees have a more negative self-concept about their competencies than men (Schmitt et al., 2021). In a similar vein, Dutta (2018) interviewed 42 Singaporean women working in STEM, who agreed that lacking self-confidence in their abilities led to organisational perceptions of incompetence, which impeded their leadership pathways.

Additionally, women who reach leadership positions in STEM experienced gender bias in the form of unequal growth opportunities and gender pay gaps (Botella et al., 2019). The participants in Dutta's study stated that even when they were given advancement opportunities, these were limited to low to midlevel management roles requiring intensive, time costly work but which tended not to lead to senior leadership positions. The gender discrimination challenge is also reported by Babalola et al. (2021), who surveyed women in STEM leadership positions in 12 African countries. These women struggled with their male colleagues who underestimated them, and unfortunately their female colleagues tended not to be supportive. This aligns with a previous study by Amon (2017) that included 46 women working in STEM fields from 11 countries, and who had experienced gender discrimination. They were not recognised as authority figures by their subordinates. They described their subordinates as not taking their directions seriously. Such gender discrimination reported by women leaders in STEM in different parts of the world might demotivate them from advancing in their career paths.

Several studies (Babalola et al., 2021; Botella et al., 2019; Dutta, 2018; Schmitt & Wilkesmann, 2020) point to work-family conflicts hindering women's opportunities to occupy leadership positions in STEM fields. They may face familial pressure and demanding jobs that impede progression in leadership roles. For instance, Dutta (2018) reported that some women viewed leadership as an additional burden on their attempts to acquire work-family balance, the latter which they valued more than becoming leaders in their STEM fields. Another reported challenge women face in STEM as leaders is the perception that they need to act 'differently' by masking emotions and changing leadership styles (e.g., Amon, 2017) or being aggressive without breaking cultural norms (e.g., Dutta, 2018). Some studies (e.g. Goris, 2020;

Minnotte & Pedersen, 2021; Watkins & Mensah, 2019) shed light on other barriers women encounter in STEM academia on their journeys to leadership positions, such as emotional and scholarly isolation and lack of departmental chair support. Such experiences decrease women's sense of belonging to their scientific communities and thus, impact their desires to seek promotions and increase turnover intentions.

Females in STEM Leadership Positions in the Arabian Gulf

The aforementioned studies were conducted in various contexts such as the United States (e.g., Amon, 2017), Europe (e.g., Botella et al., 2019; Schmitt et al., 2021), Africa (e.g., Babalola et al., 2021), and East Asia (Dutta, 2018). However, a very limited number of studies address females' presence in leadership positions in STEM fields in the Gulf Cooperation Council (GCC) region. Therefore, our present study will contribute to narrowing this gap in the literature by interviewing women in STEM leadership positions. The GCC's women's status in STEM disciplines at the university level is unique, as statistics show that females outnumber males in many undergraduate degrees in STEM majors. For instance, in the United Arab Emirates University (UAEU), which is the first public university in the United Arab Emirates (UAE), female students far outnumbered their male counterparts in all STEM disciplines in 2020 (Alzaabi et al., 2021). In Saudi Arabia, female students represent 59% in computer sciences, which is a higher rate compared to the UK and USA (Islam, 2017). In Kuwait and Qatar, females are educated to higher levels (on average) compared to males, and they dominate most STEM educational programmes, creating a reverse gender gap (Amani et al., 2021; Malallah et al., 2021). Interestingly, this has not always been the case: historically, women in the GCC region had low participation in all sectors due to conservative social norms that hinder women's roles (DeBoer & Kranov, 2017).

Unfortunately, such a high percentage of women in STEM majors in higher education institutions in GCC countries does not translate to similarly high proportions of women in the workplace. In Kuwait, Qatar and the UAE, for example, females comprise 70% of university science students, yet only 12% became part of the workforce in the science and research sectors (Islam, 2017). There are some contributing factors to the disconnection between GCC women's graduation rates and employment in STEM sectors. Culture is still an impactful factor that may hinder women's advancement in STEM professions in the GCC region, as within more traditional communities and families it may restrict women's work options to teaching, nursing, or staying at home (Al-Aawi et al., 2019; Islam, 2017). So although there might be equal opportunities for both genders in STEM careers in theory, there are social perceptions and expectations about what type of employment fits women or men (Islam, 2019). Furthermore, women might face bias in the workplace, either in the employment processes or performance evaluations (Alawi & Al Mubarak, 2019). Another influential factor, which was also reported globally, is the absence of role models who reach the top of the ladder in STEM careers (Patterson et al., 2021).

The Effect of Schooling on STEM Career Aspirations

Some studies have indicated the importance of schooling and university experiences for women to successfully continue in their STEM track, or indeed to make those choices in the first place. Sadler et al. (2012) found, through a cohort study with a large sample size, that the main predictor of a career interest in STEM at the end of secondary school was an indication of this interest at the beginning of secondary school. This would suggest that students who have STEM career interest at the point of intersection between secondary school and going on to university, are more likely to have had those interests earlier on too. Perhaps the students' experiences between these two trajectory points (the beginning and end of schooling) had helped to cement these aspirations. The effect of cultural beliefs about gender, and how these may impact on career-relevant aspirations, have shown that when students are exposed to gendered beliefs (an example of this might be that men are better at performing a particular STEM-related task), these can affect students' self-assessment of their own abilities (Correll, 2004). Therefore, the beliefs and experiences that students are exposed to at school can have a major impact on STEM career aspirations, since we know from several studies that these can be influenced by self-efficacy and self-confidence.

Examining the experiences of students who had, at one time, interest in STEM careers, but chose not to follow through with these degree choices, also shines light on the complexities of the students' decisions to follow STEM career pathways. Research underlines the effect that experiences within science communities of practice—both inside and outside of school—can have on persistence on such paths (Aschbacher et al., 2010). They explain how these experiences can "frame[d] students' perceptions of their SEM study, abilities, career options and expected success, thereby shaping their science identities and consequent SEM trajectories" (p. 564). The study also underscored the positive impact of having science advocates at school and home, and opportunities for interaction with 'real' science professionals. Receiving support in their STEM enthusiasm was also important. This did not necessarily need to be support from a STEM-related person such as a science teacher, but only someone to validate their interest and encourage them.

Students' decisions to concentrate in STEM subjects are often made during schooling, usually high school, and some research has shown that this choice is related not so much to achievement or enrolment, but instead to *interest* in science and mathematics (Maltese & Tai, 2011). As a result, Maltese and Toi discuss how policies which focus only on improving student achievement as a means to increase the number of students taking up STEM degrees may be misplaced. Other studies contradict these findings, however. For example, Wang (2013) found that achievement in high school mathematics, as well as school experiences and self-efficacy, directly influenced students' decisions to choose STEM degrees. Khattab (2015) also noted an alignment between achievement, expectations, and aspirations as being highly predictive of future educational paths.

In a largescale survey study of students in the final year of primary, and midway through secondary school, Dewitt and Archer (2015) showed that attitudes to school science were connected to science aspirations, but also that holding positive attitudes to school science did not necessarily translate into science careers. Students' involvement and participation in science activities that take place outside of school may also have a positive impact in STEM university degree pathway choices (Dabney et al., 2012). The particular science pedagogical strategies which individual teachers use can have a large impact on the interest which students have in science and the way that students perceive their teachers' attributes were also linked to their varying student interest in science (Logan & Skamp, 2013). These factors were also linked to the students' selection of science in subsequent grade levels and, it could be projected, beyond too.

Study Aims

The aims of the study are to explore the effect of the educational experiences—namely school and university—on the career journeys of successful female STEM leaders from a sample of GCC countries. In particular, what events, situations or influences during these educational years had an impact on the women's successful STEM career journeys?

The study aimed to answer the following research questions:

RQ 1. What were the schooling experiences of female STEM leaders in the GCC, and how did this impact upon their perceptions of STEM and their journeys to become STEM leaders?

RQ 2. How did the university education experience, including overseas experiences, impact upon their ability to persist and succeed, and what skills were acquired during this time?

Methodology

The study participants were female STEM leaders from the following GCC Countries: UAE, Oman, Kuwait, Bahrain and Qatar. We initially adopted non-probability purposive sampling methods to recruit our sample, with suitable participants asked to satisfy the following criteria:

- GCC citizens of the named countries
- Currently holding a leadership role within the included GCC country in a STEM related field (which could be either corporate or higher education)

We made use of professional networking social media sites, along with our network of researchers and colleagues to identify and invite potential participants in academic, corporate and/or industrial STEM settings. We also utilised snowball sampling methods by asking participants to forward the invitation to female colleagues who met the study criteria. Participants were invited to take part in the research study by an initial email, which contained some brief information about the study. Interested participants were then followed up to arrange an interview time convenient to them. The interviews took place online partly due to the geographical locations of the participants, and also due to the fact that data collection took place during the COVID-19 pandemic. We were originally aiming to recruit approximately 13–15 participants and managed to recruit the upper figure of this projection (fifteen). For in-depth individual qualitative study such as this, this can be considered an adequate number of participants (Marshall et al., 2013).

Qualitative data collection was collected by way of interviews which took place on a one to one basis, due to the possibility of sensitive and confidential data arising from the participants' narratives. The framework underpinning the interview questions was devised from the key concepts which arose from the literature review, however, a semi-structured style of interviewing was used whereby the interviewer was guided by the interview questions, yet still had the flexibility to take a participant's lead if they wished to elaborate more on particularly resonating themes, and to ask follow up questions too. This style of interviewing tends to generate richer data sets than structured interviewing techniques (Adams, 2015).

The interview comprised of fourteen questions, and each interview took between 40 min to one hour to complete. The participants were offered the opportunity to carry out the interview in either Arabic (the mother tongue of all of the participants), or in English, but all opted to participate in English. This was presumably due to their own fluency and comfort level in communicating in English due to having studied in this medium for many years. The interviews were audio recorded with the express permission of the participant, and later transcribed by a research assistant.

The question structure provides a foundation with which to begin coding emerging themes. After the initial main themes emerged, and data classified according to these, sub-themes were then developed in sequential analysis. The themes arose within the research question domains—significant experiences of schooling such as interactions with teachers, peers, learning experiences, etc. The general approach to coding data described by Miles and Huberman (1994) was adopted. In this, a list of codes connected to the interview questions is developed prior to data collection. These are based on the conceptual framework which arises from the literature. Codes are then be kept, adapted or discarded for their fit (or lack of) to the data. As multiple interviews are analysed, more codes may be added as they emerge from the data. From these patterns, we were then able to identify broader themes. These themes could be induced according to relevancy to the research questions (Robson, 2011).

We were very mindful of the ethics of the research study, and the fact that the chief potential risk is in the sensitivity of the subject matter itself. The questions were pertinent, and at the same time, highly personal, so we were wary of taking care of the risk of invoking discomfort. We reassured participants of their right not to answer any question they did not wish to, or indeed to leave the study at any time with no justifications required. Assurances of confidentiality were given both in writing (in the consent form and information document) and at the beginning of the interview itself. Identifying factors were carefully disguised in reported data, since the explicit reference to their particular employment or details of the specifics of their institution might easily identify them. In order to facilitate the trustworthiness of the participants' responses, the schedule of interview questions was carefully critiqued by a volunteer in a senior STEM leadership position, who would not partake in the study. We also endeavoured to ensure that participants were able to speak freely in a way which would be most likely to garner data which was authentic. The environment was carefully nurtured to enable the participant to feel relaxed, and to generate the most truthful data possible, and transcripts were also sent to the participants to check for their accuracy in representing the interview dialogue.

Findings and Discussion

We now present the key findings of how education has impacted upon the participants' decision to follow a STEM career, and how might this have affected the women's career path.

Schooling Experiences

There was some evidence in the interview data that the participants' schooling experiences had impacted on their early perceptions of science and STEM subjects generally. Experiences of interactions with these subjects, and their teachers, were extremely mixed. Key themes which arose include some fairly negative interactions with teachers, but these were also interactions which helped the women to develop resilience and determination to succeed in any case. In addition, while the home and family experiences of the women were diverse, those with highly educated family members already working within STEM fields talked about the expectations placed upon them to follow a similar career path. This was sometimes perceived as useful, and sometimes not.

An interesting finding from the interviews was that participants frequently referred to their interest in STEM subjects as having developed over time, and for some, not until they were already enrolled in their university degree course. Instead, some had chosen their STEM-related degrees due to more pragmatic reasons, such as having shown aptitude in the subjects, and the experience of having gained high marks in those subjects, along with gaining the recognition of teachers who then encouraged them to choose STEM-related subjects and tracks in upper high school, rather than choosing the path due to a genuine love or interest for the subjects, which came later. One participant explains, she was: "good at the subjects, perhaps more than having a love for the subjects".

The participants who had attended government schools referred frequently in discussions of their schooling years to the science 'track', which was a route selected for students to pursue at around grade 10 level. Generally, only students who had performed well in STEM subjects in examinations prior to this, would be able to select this track. A major challenge mentioned by participants was the need to learn in English, as was (and still is) the policy in universities in the region, but definitely if students wished to attend overseas universities, proficiency in English language was a major consideration and a hurdle to overcome for some. This mismatch between the language medium in which they had learned science at school, which was Arabic for those attending government schools, and then switching to English later, was a challenge for many participants, particularly where it was felt that the methods of teaching English at school had been far from ideal. One participant explains how she:

> had to take some English classes, I had to develop my English skills as well. It is not like I don't understand the theory behind things, I understand them very well, but I cannot communicate to my professors everything .... I used to get Bs and Cs just because I don't understand the question.

The idea that particular individuals in schools, such as teachers and administrators, could motivate or encourage individuals into following STEM paths, was a strong theme in the participants' responses. One participant describes a mathematics teacher, with whom she still has a strong relationship to this day:

> there used to be a good teacher we all loved, and she taught me a lot of classes, math classes, so I think this has an effect also because the teacher sometimes drives you to your interest in a good way.

Generally, teachers who made the students feel good about themselves and paid attention to them as individuals, appeared to have a large impact on the participants. There were references to enjoying the way particular teachers delivered classes. One participant described her experience of attending a private school in the UAE, feeling that she was in the minority (in her nationality) among expatriate children, and was treated differently by teachers as a result. She felt that her teachers had a very traditional perception of life for Emirati women. She stated that:

> Emirati women are not going to use their education, yes nice to have it for them. They probably get married when they finish high school. So, I don't think my teachers have a lot of faith in me. To be honest with you in terms of actual support in the schools, you know they like oh, they're rich, and privileged and they're going to get married and stuff like that.

This type of stereotyped perception of Gulf girls and women seemed to work either to alienate the individual from school and from science classes, or to have the opposite effect; to cause women to want to prove themselves further, and to prove the teacher wrong. Some cited this as a challenge which actually pushed them forward and motivated them to show the teacher how wrong she was. The participant who gave the citation above explained that the teachers' prejudice did not "stop my career moving forward". However, it did create an additional barrier for the women (then girls) to get through. Another participant, also educated in the private school sector, described how she felt that she and a fellow student were:

> treated as outcasts by the school basically for insisting on following the STEM track, and we did IGSCEs. I think they were worried we would bring down the school reputation or something. But my friends and I liked to break the mould!

The somewhat tenacious attitude portrayed here is fairly typical of participant responses too, in the face of such situations.

There were also references to school competitions, both within schools and externally, which some participants considered to be motivating. This is, of course, something which divides opinion, but in the case of a few participants, was considered to be constructive rather than divisive. Others found the competitive nature of schooling to be detrimental to them, and to cause them to feel disengaged from the classroom environment, since they felt the focus to be only on grades, and:

> on students who are having high marks … the rest they don't exist unfortunately … I have not the best experience with the schools … [if] there is a math competition or physics competition I will not be chosen because I wasn't on the top five students in the class or whatever in the school so I was not chosen that puts me down a little bit.

An interesting theme which arose was that often, getting to university seemed to be a goal in itself, rather than specifically studying a STEM degree when they got there. This was usually explained with reference to family expectations, or a sense that they were contributing to their country's capacity building by increasing the social capital within that country. Interestingly, this was true for women who had families with high levels of education, but also equally true for those who did not. Indeed, half of the participants mentioned that at least one of their parents was either illiterate, or not educated beyond primary level. Many women talked about how their close family relatives (mostly males) who had pursued STEM careers had encouraged them as girls to develop an interest in these subjects. They did this by taking them to their workplaces, talking to them about their own careers, etc. For women in these families, there was a strong expectation that they too would pursue university degrees and forge strong career paths, as one participant's reflection shows: "the focus was not thoroughly on science, but the sky was the limit at that stage".

For some women, particularly those whose families came from STEM backgrounds, it was felt that a STEM career was an expectation:

> my parents, their backgrounds are in STEM, so they encouraged me. [they] both did grad school, they are both STEM and they had the expectation and very much encouraged me to pursue STEM and pursue a graduate degree in STEM and get a PhD in STEM. So, the raw energy came from them

For others, there were pivotal moments with either specific teachers, or particular learning experiences, which sparked interest in the subjects. One participant described very animatedly that an experience with a science teacher had stayed with her throughout adult life, as evidence of how students can be deeply impacted by a learning experience. She explains how the experience "really ignited things for me, and I still remember that till today." The participant vividly remembered a school trip as a primary aged child, where her science teacher had taken her to the sea and pointed out various geological features and species of fish and birds, and inspired in her a great love for nature which led to a career in STEM. This story is important, and underlines the value and significant impact which teachers can have on their students, sometimes unknowingly.

Impact of Higher Education, Including University Experiences

Key themes arising in the analysis of higher education experiences include parental support for education, and some family resistance around international travel. A male relative was often mentioned either as a travel companion or mediator, as was in keeping with Islamic traditions (McIntosh & Islam, 2010) and which often alleviated this family concern.

All participants but one studied for at least a portion of their higher education (mostly graduate studies) overseas in international universities. For older participants in particular, this often happened at a time when overseas travel for women from the Gulf was still relatively unusual. Some of the women were accompanied, at least at the beginning, by a male relative, while some travelled completely alone. All of the participants described the importance of having family support (particularly from fathers) to enable them to do this, and also that they had sometimes needed to argue and negotiate with their families to be permitted to study abroad. However, many fathers are described as being not only supportive, but instrumental to the women's ability to study abroad. One participant even remarked that when an elder brother had questioned her decision to travel, her father had stepped in to declare it "none of the brother's business", which was appreciated. Others described their parents as encouraging them to go, and being more "forward-thinking" about them receiving that opportunity. One participant described a conversation with her father wherein he said that:

> you are studying for yourself, and your study is your weapon. One day I will pass away and no one will be for you, only your study no one will help you ...

The overseas experience was not without its challenges. The women spoke of the terrible isolation and alienation that they felt at times in a new culture and being far from home. Some also felt that they were perceived differently from other students, being both female and from the Middle East, as this statement demonstrates: *At that time, it wasn't easy for a woman to leave the country and go abroad. Even in the States, everybody was looking at me like "Where did you come from? Where is your country?"*

For those who had taken up graduate study in earlier decades (1980s to 2000s), communication was also a major challenge, and some women described their weekly trips to payphones to make a quick call to their families, pre-internet days. One woman poignantly described the experience:

> I'm telling you the first three months, it was a nightmare, I used to cry, and I was like why did I come here, why did I choose all of this? I was homesick, and you know, you had been

raised with a big family, and then you suddenly leave on your own, so it's difficult with a different culture, and I didn't know anyone ....

Motivations to study abroad included having had some interactions with faculty in their undergraduate study experiences, who had either studied in those institutions themselves, or who had articulated what one participant described as having determination to join "the best university in the world" to do graduate studies. Others were encouraged by senior administrative staff, such as their academic dean, to travel abroad and study. For many participants, the main attraction of studying abroad was academic, for example, the woman who felt that:

the environment was purely scientific ... you have many people to talk to about science ... the interaction was so great compared to what I had earlier, then I really appreciated much more

indicating a level of challenge and vigorous discourse around science that perhaps she had not experienced previously.

For some, the turning point of their initial homesickness, culture shock and challenges in acclimatising to their new country of residence, came when they were able to establish micro-communities of other international students with whom to socialise and provide mutual support. These were usually other women from the Gulf, and indeed a great camaraderie and community spirit appeared prevalent here. Soon, the women were also able to help other newcomers, as this participant describes:

The office of international students started connecting us with the Saudi girls, the Kuwaiti girls, you know. Just to feel like you are not on your own here, people from your place they are here. We used to meet every Saturday, once in a month. Then I became like - anyone who is coming new - they will ask me to come and talk with them, and talk about their experience about the culture shock, and try to help them

Important religious festivals were particularly difficult times for the women to be far away from their families, such as at Eid, or during Ramadan. Women who had selected to study in countries nearer to the region such as the UK, had struggled slightly less with this, as they were able to travel home more easily compared to peers in North American universities. This choice of location was a frequent point of discussion in the interviews, with one participant describing it as changing upon marriage, i.e.

the boundary is Arab countries (laughs) so I was allowed only to go to [city] you know, while my brother went to USA but when I went to a postgraduate degree I was married already so I was like accompanying my husband and I took advantage of that

The married females often had young children by the time of their graduate studies, and relayed their experiences of combining motherhood with their studies. While there were challenges involved in juggling these two roles, their children also brought joy and a sense of balance to them. Some women had spouses who regularly visited or spent prolonged periods of time with them and helped with childcare during those times. The support of families, including spouses, was considered intrinsic to the women's journeys:

> You will have to choose either family or career, but if you want to have it all, if you want to have family and a career, you must have family support. Without it is impossible.

Where there was an option to study abroad without the children, leaving them at home with spouses or family, participants chose not to do this. One participant explained how she was "totally against that. I feel either I can do it all or I wouldn't do it". Another described taking her two very young children with her to the lab and a colleague commenting kindly that she was like a 'duck' with two little ducklings walking behind her to the lab. For her, having her children with her overseas was a positive affirming experience which she felt supported in by her colleagues.

A prevailing theme from the testimonies of the women was the skillset which they developed as a result of their overseas experience, quite separate from the content knowledge and skills which they learned during their degree. Key was the sense of developing resilience and becoming independent, which was described by one woman as "a great experience; it added a lot to my personality". The idea of facing and overcoming fear, was frequent, and learning to depend on oneself was a major outcome of the experience. It was felt that those kinds of skill developments were very transferable and useful to their STEM career later, and indeed many participants did refer to that kind of resilience building and learning to deal with challenging times, as being helpful in the STEM workplace. Others described the sharp rise in self-esteem and self-efficacy they experienced due to having successfully completed their degree, and how their previous sense of 'imposter syndrome' had lessened because of this:

> There were many instances where I felt as if I wasn't good enough in the position, and the job I was given. I think that it drove me to work extra hard to prove that I was a good candidate. I think that it lessened a lot when I finished my MBA, so a stamp of approval from an Ivy League University is what gave me that feeling of competence.

The fact that the overseas experience gave the women an opportunity to explore different cultures was important, in broadening their viewpoints and diversity of ways of thinking, as this statement illustrates: "learning how people live, or how you eat! I felt like I never studied English, like, what's going on here, why can't I understand anyone?" For some, having to interact with students with such variance of viewpoints from their own, strengthened their own thinking and ability to argue, such as the woman who reflected: "I explored myself there, I knew who I am. It was very difficult but lots of joy".

One participant reflected on the growth of online learning opportunities for international students nowadays. She felt that while it had benefits in terms of increased access, perhaps increasing accessibility to women who would not otherwise be able to study, its detriment was that they were missing out on subtle interactions which took place outside of formalised learning time. One particular aspect of the international university experience, which was new to some, was the exposure to the world of competitive grant submissions, which one participant felt was "something completely different from the MENA (Middle East North Africa) region". This exposure could very foreseeably have been a useful skill when they returned to their home countries, either within STEM industry or in an academic setting. One woman explained just that, and stated that her current success with grant awards came from the experience that she gained while overseas. The experiences, while filled with learning and exposure to critical thinking skills, were inspiring and challenging in equal measure. We have already explained how the isolation was at times overwhelming for women. In addition to this challenge, some reported facing prejudices which were grounded in stereotypes about them both as women in a STEM context, and as women from the Middle East. One woman described a particularly upsetting incident whereupon she had gone into a computer lab to work with her class, and although the whole class were working alone, she was explicitly instructed not to work alone, but to find a partner. Her instructor told her to "work with someone else because we don't want you to break the computers". She described how upset she was, and how in that moment, she thought of how accomplished she was in computing skills, and the offence she felt:

> I was like okay well they think I came from a tent, that's the media what they are trying to show, and now he believes that I don't understand how to deal with computers

Role Models in Education

Although this was not a theme inherent in the research question, it arose organically from the data and links to the literature on the importance of girls and young women having role models in STEM. There was evidence through the participant responses that some were particularly affected by teachers whom they experienced as role models, both teachers in school and faculty in higher education. The genders of the role models were both male and female, though at the university level, this was predominantly male. This is to be expected given the predominance of males in those fields in higher education, particularly at that time in the Gulf, for older students. In schools, due to the gender segregation policies operating in most public schools in the Gulf, most of the teachers were female. There were some examples of role models due to their personal characteristics which were lauded, such as one participant who said that they found their headmistress to have acted as a kind of role model. This was due to them being 'very encouraging', and another who was inspired by a teacher who 'drove her interest'. Many role models were non-nationals too, again, quite typical of the make-up and demographic of educators at that time.

Many of the participants referenced maintaining long-term relations with teachers from university, some of whom had nurtured and encouraged them to continue in their STEM career journey. For some, faculty who had particular notoriety in their roles and fields were significant to them, such as one participant who described her lecturer as talking about things that "no one else was talking about at that time", and being a "well-known face, a celebrity". She felt that he had guided her towards a particular specialism field in science.

One participant said she thought (in the 1970s and 1980s) that anyone with a PhD was a role model to her. Another recounted meeting someone who she considered a role model, thinking that she was:

> the first woman that had PhD, I looked at her, you know, wishing that I will be like her one day … just to introduce herself as doctor, and this field is specialized, and she's working in it!

For the participants working in academia, a high-ranking female was a clearly symbolic role model, showing them that such a position was possible for them too, which had a large impact on them. Speaking of one faculty member, one participant explained:

> She was a full professor, so when she asked me - what do you want to be when you finish your bachelor? I told her I want to be a professor, I want to be lecturing in academia, so this was a role model. I was looking up to her

A further theme in connection with role models is that their particular style of teaching enticed the participant into sustaining a passion for their STEM subject, such as the way that they engaged students, their way of 'making meaning' of learning, and making this accessible and relevant to them. This is exemplified in the following excerpt:

> Even the equation of physics he taught us actually how to sense the equations, and it's not only numbers and letters, we should think about it as it's real as a language, and he was encouraging us to create

One participant spoke about a teacher who "inspired us and made us love physics … he was funny at school", and gave examples of humorous examples the teacher had given. That particular faculty member also asked students to think about their motivations for being on their degree course, to think beyond this and really look at what inspired them. As she described, this teacher was to her "a role model; he is a very, very, very unforgettable person".

Implications of the Study

In the UAE, the National Vision 2021[1] highly emphasises the participation of both female and male national citizens in higher education and ultimately to participate in the national workforce. This workforce, it is hoped, will be skilled enough to contribute to the growth of the future UAE Knowledge Economy, in which STEM plays a large part. The UAE government, and the legacy laid forth by Shaikh Zayed Al Nahyan, ensures that Emirati women have been given many opportunities within both policy and practice to succeed in their education and in the workplace; indeed there are many examples observed frequently in public media of Emirati women who have excelled in STEM careers in the UAE. Other Gulf countries have created similar opportunities in STEM fields to women, to varying degrees. The 'leaky pipe', 'gender filter' and 'glass ceilings' (Clark Blickenstaff, 2005; Glass & Cook, 2016) whereby numbers of females in STEM careers leave soon after beginning, or are

[1] https://www.vision2021.ae/en.

retained yet remain in middle management positions, is as much of an issue in the GCC as internationally. Research into the motivations of those women who manage to bypass both phenomena is therefore timely.

The key findings of the study have implications for all of the GCC countries, and could impact on long-term recruitment and retention of women working in STEM fields. While the limitations of the study include the fact that the participants are of course only a sample of female STEM leaders, they are from different Gulf countries, and work across a variety of STEM fields. The broad themes that arose from the analysis of their responses would suggest that they could potentially be generalised to other female STEM leaders.

In summary, the key findings and their potential implications, are as follows:

1. **Interest in STEM was not necessarily the reason for university choices**

An interesting finding from the interviews was that participants frequently referred to their interest in STEM subjects as having developed over time, and for some, not actually until they were already enrolled in their university degree course. High achievement in STEM-related subjects was often cited as a reason for choosing a STEM degree, rather than a genuine love or interest for the subjects, which came later. An implication of this finding could be that it might be beneficial to offer advisory sessions for high school students in grades 11 and 12, who might be potential STEM undergraduates, and perhaps introducing schemes to shadow women working in STEM fields. Through mechanisms such as these, students will be introduced to STEM paths they might take at the university level, either inside the country or overseas. Importantly, given the finding, it would be critical to offer all of these options to all girls, not only to those who had previously expressed interest in STEM subjects.

For many of the women, getting to university was a goal in itself, rather than specifically going to study a STEM degree. Again, advisory sessions would be useful to provide students with in-depth information about possible STEM majors for the undergraduate level.

2. **Individual teachers and administrators often act as key motivators**

The idea that particular individuals in schools, both teachers and administrators such as school principals, could motivate or encourage individuals into STEM paths, was a strong theme in the participants' responses. Generally, teachers who made the students feel good about themselves and paid attention to them as individuals, appeared to have a large impact on the participants. There were references to teachers' delivering classes well, which they described as leading to their enjoyment. Based on this finding, teachers' and administrators' roles as a source of inspiration and encouragement should be emphasised during school personnel professional development (PD) sessions. Awareness in teachers and administrators should be raised about the fact that their teaching practices may impact students' decisions to pursue a career in STEM. Again, one possible way to increase this awareness is through well-designed PD sessions. Another implication is the necessity to enhance teaching quality, especially in STEM classes, as it influences students' future choices regarding undergraduate majors. This is indicated in the participants' responses, who linked enjoyment of STEM subjects and lessons to what they perceived as teaching quality, which affected their attitudes towards STEM subjects.

3. **Experience of prejudice from school teachers**

The stereotyped perception of women from the Gulf, which we observe in some of the participants' responses, seemed to work either to alienate or to strengthen their resolve to persist in STEM. This finding shows the importance of protecting female students from such prejudices regarding their science abilities. School environments should be a place where students' abilities are maximised instead of being limited or discouraged by teachers. Policies and practices in schools could be developed so that students could report and discuss any prejudice they experience to a neutral figure in schools, such as social workers or advisors. Also through discussions and mentoring, students can be taught to develop coping mechanisms and response strategies to overcome these experiences, which might otherwise hinder their passion for STEM subjects.

4. **Pivotal learning moments**

For others, there were pivotal moments with particular learning experiences, which ignited a flame of interest in the subjects. An implication for this finding is the importance of designing rich and meaningful learning experiences in the STEM classroom for students. Such learning experiences can spark the passion for science subjects among students and encourage higher numbers of females to take up STEM careers, and hopefully, reach leadership positions. Particular styles of teaching also led to the participants sustaining a passion for their STEM subjects, such as in the way that teachers were perceived to engage with students, their way of 'making meaning' of learning and making the learning generally accessible to them. This finding supports implications such as the importance of enhancing teaching quality and designing rich learning environments.

5. **Challenging overseas experiences**

The women spoke of the initial isolation of their overseas experience. This was alleviated by the establishment of micro-communities with other international students. These were usually other women from the Gulf, and indeed a great camaraderie and community spirit appeared prevalent here. These experiences by the majority of female STEM leaders in this study indicate the value of establishing support networks through students' clubs, events and trips. Since the indicators from our data is that this was one of the key reasons for the women persisting with their studies, it should not be under-valued as a vital mechanism in the retention of females in these fields.

6. **The appeal of challenging academic environment**s

For many participants, the attraction of studying abroad was academic, for example, the woman who felt that "the environment was purely scientific … you have many people to talk to about science … the interaction was so great …", and the academic vigour she experienced and relished. This finding may indirectly highlight the lack of a strong academic environment in the Gulf region, which might contribute to women leaving STEM sectors or perhaps not choosing to enter those fields in the first place. This implication needs a more detailed consideration than is possible in this chapter, but it certainly appears to suggest that focus should be directed towards enhancing the quality of academic environments in the Gulf region.

7. **Juggling career with motherhood**

While there were challenges involved in juggling parenting and study commitments, the women's children also brought joy to them during their times of study overseas. The implication of this is that women should be supported through policy to be able to do this, for example with the provision of free or subsidised childcare at the institutional sites, as some universities do have. Some women had spouses who regularly visited and supported with childcare during those times, but not all. So, the women found childcare centres very useful in helping them to manage motherhood responsibilities while studying abroad, which they missed in their home countries. If women are supported to reach an effective balance of work and home by mechanisms such as the opening of these childcare facilities, it is very likely to help with the retention of females in STEM degrees of study.

8. **Developing resilience**

A prevailing theme from the testimonies of the women was the development of personal skills such as resilience and independence, which they developed as a result of their overseas experience. This was something quite separate from their development of academic skills such as content knowledge. This was powerfully described by one participant as being "a great experience; it added a lot to my personality". This finding shows that studying overseas is a broader experience than merely academic content acquisition, and one which has the potential to shape female STEM leaders' personalities and develop their personal qualities such as strength and tenacity, which are essential in the STEM workplace. This arises as a result of a multitude of factors, including meeting a diversity of students and learning to acclimatise to a new culture. In the Gulf region, the last two decades have seen a mushrooming of the numbers of international branch campuses from well-established universities, mostly from the US, UK and Australia. While these campuses offer an experience which may be more diverse in terms of teaching faculty, curricula and of course offer the advantages of being able to access this type of education in one's home country, they do not offer the complete immersion into a new country and cultural setting which the overseas experience offers, along with all its related challenges. So an implication of this finding might be that countries in the Gulf should continue to promote studies in overseas institutions, including scholarship programmes, rather than perceiving less need for these. Having said that, not all females in the Gulf region are able to study abroad for various reasons such as family commitments or social constraints. The point here is that women need to be able to access a number of alternatives, including offerings of short and extensive courses overseas which offer greater flexibility. Access to this variety of higher education is paramount to an agenda of recruitment and retention of female STEM candidates.

9. **Teachers as Role Models**

There was evidence that some of the participants were particularly positively affected by teachers whom they experienced as role models, both teachers in school and faculty in higher education. The genders of the role models appeared to be irrelevant. An implication of this finding is simply that role models *are* important to some women. Higher education institutions should expose students to successful female STEM leaders in their countries by organising themed conferences and workshops, connecting students with potential role models. For the participants working in academia, a high-ranking female was a clear role model, showing them that such a position was possible for them too. This finding indicates that enabling women to reach leadership positions will encourage other females to believe in their abilities, and that for younger women, being able to visualise this was helpful and inspirational to their own journeys. Policies to encourage this could include the use of quotas, whereby top management stakeholders in both academic and industrial sectors specify a pre-determined portion of leadership positions to be women. Such a policy may allow upcoming generations of female STEM professionals to have role models they can relate and aspire to.

Conclusion

In this chapter, we have recounted the educational experiences of female STEM leaders, and described ways in which these experiences may have influenced their decisions to pursue STEM career paths. In school, the main influences have been individual teachers who motivated the women, either through positive and encouraging interactions with teachers, through experiencing engagement and interest in their lessons, and these influences have been intertwined with families' encouragement of them to take interest in STEM subjects at school. Several women experienced negative expectations from their schoolteachers, which was difficult yet gave them a 'push' to try to 'prove them wrong'. Articulating particular interest in STEM subjects was not a prerequisite for deciding to take up degrees in this field, but most of the women in the study had academically performed well at school and so may have been receiving strong encouragement from teachers and families to follow a STEM path as a result. In later university studies, the women were influenced by teachers who stood out to them for one reason or another, particularly where the teacher took an interest in their studies and offered career advice. Most of the women had undertaken graduate studies overseas, an experience which was challenging, inspiring and character building. Schools, universities and educational bodies liaising with industry to recruit and retain females in STEM careers can implement some of the recommendations to practice outlined in this chapter in order to make use of the exploration of the 'experts' in the field, i.e. the women who already surpassed the challenges and have much to teach the younger generation of women coming up into STEM career pathways.

References

Adams, W. C. (2015). Conducting semi-structured interviews. *Handbook of Practical Program Evaluation, 4*, 492–505.

Al-Aawi, A., Elias, H., Zaid, F., Alroaili, M., & Al-Bassam, S. (2019). Factors affecting Bahraini women working in the engineering fields. *Journal of International Women's Studies, 20*(7), 289–318.

Alawi, W. S. S., & Al Mubarak, M. S. (2019). Gender gap in science, technology, engineering and mathematics: Barriers and solutions. *International Journal of Economics and Financial Issues, 9*(6), 225–231. https://doi.org/10.32479/ijefi.8908.

Alzaabi, I., Ramírez-García, A., & Moyano, M. (2021). Gendered STEM: A systematic review and applied analysis of female participation in STEM in the United Arab Emirates. *Education Sciences, 11*(573), 1–14. https://doi.org/10.3390/educsci11100573.

Amani, S., Youssef, E. M., Alagha, R. Y., Hillman, S., & Ruimi, A. (2021). "This is a very male job": Challenges encountered by females during recruitment and hiring for engineering jobs in Qatar. Paper presented at the *ASEE Virtual Annual Conference Content Access*, Virtual Conference. https://doi.org/10.18260/1-2--36538.

Amon, M. J. (2017). Looking through the glass ceiling: A qualitative study of STEM women's career narratives. *Frontiers in Psychology, 8*(236), 2–10. https://doi.org/10.3389/fpsyg.2017.00236.

Aschbacher, P. R., Li, E., & Roth, E. J. (2010). Is science me? High school students' identities, participation and aspirations in science, engineering, and medicine. *Journal of Research in Science Teaching: The Official Journal of the National Association for Research in Science Teaching, 47*(5), 564–582. https://doi.org/10.1002/tea.20353.

Babalola, O. O., du Plessis, Y., & Babalola, S. S. (2021). Insight into the organizational culture and challenges faced by women STEM leaders in Africa. *Social Sciences, 10*(105), 2–16. https://doi.org/10.3390/socsci10030105.

Botella, C., Rueda, S., López-Iñesta, E., & Marzal, P. (2019). Gender diversity in STEM disciplines: A multiple factor problem. *Acta Paedagogica Vilnensia, 21*(1), 1–17. https://doi.org/10.3390/e21010030.

Clark Blickenstaff, J. (2005). Women and science careers: Leaky pipeline or gender filter? *Gender and Education, 17*(4), 369–386. https://doi.org/10.1080/09540250500145072.

Correll, S. J. (2004). Constraints into preferences: Gender, status, and emerging career aspirations. *American Sociological Review, 69*(1), 93–113. https://doi.org/10.1177/000312240406900106.

Dabney, K. P., Tai, R. H., Almarode, J. T., Miller-Friedmann, J. L., Sonnert, G., Sadler, P. M., & Hazari, Z. (2012). Out-of-school time science activities and their association with career interest in STEM. *International Journal of Science Education, Part B, 2*(1), 63–79. https://doi.org/10.1080/21548455.2011.629455.

DeBoer, J., & Kranov, A. A. (2017). Key factors in the tertiary educational trajectories of women in engineering: Trends and opportunities in Saudi Arabia, the GCC, and comparative national settings. In A. Siddiqi & L. D. Anadon (Eds.), *Science and technology development in the gulf states* (pp. 56–88). JSTOR.

DeWitt, J., & Archer, L. (2015). Who aspires to a science career? A comparison of survey responses from primary and secondary school students. *International Journal of Science Education, 37*(13), 2170–2192. https://doi.org/10.1080/09500693.2015.1071899.

Dutta, D. (2018). Women's discourses of leadership in STEM organizations in Singapore: Negotiating sociocultural and organizational norms. *Management Communication Quarterly, 32*(2), 233–249. https://doi.org/10.1177/0893318917731537.

Fatourou, P., Papageorgiou, Y., & Petousi, V. (2019). Women are needed in STEM: European policies and incentives. *Communications of the ACM, 62*(4), 52. https://doi.org/10.1145/3312565.

Glass, C., & Cook, A. (2016). Leading at the top: Understanding women's challenges above the glass ceiling. *The Leadership Quarterly, 27*(1), 51–63.

Goris, T. (2020). Where are they? Limited female presence in STEM professions. *The Midwest Quarterly, 61*(3), 330–336.

Islam, S. I. (2017). Arab women in science, technology, engineering, and mathematics fields: The way forward. *World Journal of Education, 7*(6), 12–20.

Islam, S. I. (2019). Science, technology, engineering and mathematics (STEM): Liberating women in the Middle East. *World Journal of Education, 9*(3), 94–104. https://doi.org/10.5430/wje.v9n3p94.

Khattab, N. (2015). Students' aspirations, expectations and school achievement: What really matters? *British Educational Research Journal, 41*(5), 731–748. https://doi.org/10.1002/berj.3171.

Logan, M. R., & Skamp, K. R. (2013). The impact of teachers and their science teaching on students' 'science interest': A four-year study. *International Journal of Science Education, 35*(17), 2879–2904. https://doi.org/10.1080/09500693.2012.667167.

Makarem, Y., & Wang, J. (2020). Career experiences of women in science, technology, engineering, and mathematics fields: A systematic literature review. *Human Resource Development Quarterly, 31*, 91–111. https://doi.org/10.1002/hrdq.21380.

Malallah, S., Alfailakawi, S., Alkhurafi, T. Y., & Weese, J. L. (2021). Reversing gender stereotypes in STEM education in a gender-segregated region. Paper presented at the *ASEE Virtual Annual Conference Content Access,* Virtual Conference. https://peer.asee.org/37682.

Maltese, A. V., & Tai, R. H. (2011). Pipeline persistence: Examining the association of educational experiences with earned degrees in STEM among US students. *Science Education, 95*(5), 877–907. https://doi.org/10.1002/sce.20441.

Marshall, B., Cardon, P., Poddar, A., & Fontenot, R. (2013). Does sample size matter in qualitative research? A review of qualitative interviews in IS research. *Journal of Computer Information Systems, 54*(1), 11–22. https://doi.org/10.1080/08874417.2013.11645667.

McIntosh, J. C., & Islam, S. (2010). Beyond the veil: The influence of Islam on female entrepreneurship in a conservative Muslim context. *International Management Review, 6*(1), 102.

Miles, M. B., & Huberman, A. M. (1994). *Qualitative data analysis: An expanded sourcebook.* Sage.

Minnotte, K. L., & Pedersen, D. E. (2021). Turnover intentions in the STEM fields: The role of departmental factors. *Innovative Higher Education, 46*(1), 77–93. https://doi.org/10.1007/s10755-020-09524-8.

Patterson, L., Varadarajan, D. S., & Salim, B. S. (2021). Women in STEM/SET: Gender gap research review of the United Arab Emirates (UAE)—A meta-analysis. *Gender in Management: An International Journal, 36*(8), 881–911. https://doi.org/10.1108/GM-11-2019-0201.

Robson, C. (2011). *Real world research* (3rd ed.). Wiley.

Sadler, P. M., Sonnert, G., Hazari, Z., & Tai, R. (2012). Stability and volatility of STEM career interest in high school: A gender study. *Science Education, 96*(3), 411–427. https://doi.org/10.1002/sce.21007.

Schmitt, M., Lauer, S., & Wilkesmann, U. (2021). Work motivation and career autonomy as predictors of women's subjective career success in STEM. *Acta Paedagogica Vilnensia, 46*, 73–89. https://doi.org/10.15388/ActPaed.2021.46.5.

Schmitt, M., & Wilkesmann, U. (2020). Women in management in STEM: Which factors influence the achievement of leadership positions? *International Journal of Gender, Science and Technology, 12*(3), 328–352.

Wang, X. (2013). Why students choose STEM majors: Motivation, high school learning, and post-secondary context of support. *American Educational Research Journal, 50*(5), 1081–1121. https://doi.org/10.3102/0002831213488622.

Watkins, S. E., & Mensah, F. M. (2019). Peer support and STEM success for one African American female engineer. *The Journal of Negro Education, 88*(2), 181–193. https://doi.org/10.7709/jnegroeducation.88.2.0181.

Wilkins-Yel, K. G., Bekki, J., Arnold, A., Bernstein, B., Okwu, C., Natarajan, M., & Randall, A. K. (2021). Understanding the impact of personal challenges and advisor support on STEM persistence among graduate women of color. *Journal of Diversity in Higher Education, 1–14.* https://doi.org/10.1037/dhe0000236.

Chapter 9
Gender Representation in STEM Departments in Higher Education Institutions in the UAE

Martina Dickson and Masada Al Harthi

Abstract Inequitable representation of females in professorial roles in universities is well documented internationally, and explained through the use of interesting metaphors such as the 'leaky pipe' or the 'glass ceiling'. Documented reasons for this include the fact that women tend to shoulder most caring responsibilities, are most impacted by leave periods to have and take care of young children, and are less likely to be promoted, all else being equal, into terminal academic ranks. In the Arabian Gulf country of the United Arab Emirates (UAE), the situation for female academics is slightly different, in theory, and caring responsibilities may be less contentious due to a plethora of availability of inexpensive childcare and domestic help. Among citizens of these countries, there are also various policies of support that impact on women's capacity to access education and academia. In this chapter, the status of women in academic professorial and leadership roles in STEM departments and in higher education generally in the United Arab Emirates, is examined in order to draw comparative indications of female representation in higher education. We argue that all evidence indicates that in fact, organisations remain highly gendered, with male professors dramatically outnumbering females across all key universities in the country, and that the reasons for this underrepresentation are more complex than may appear.

Keywords Female · Representation · Higher education · Leadership · United Arab Emirates

Introduction

The number of females studying in higher education has risen dramatically over the past few decades so as to often outnumber males in some degree programmes. This includes programmes that were traditionally male-dominated, such as STEM degree programmes, and increasingly, women make up equal proportions of the

M. Dickson (✉) · M. Al Harthi
Emirates College for Advanced Education, Abu Dhabi, UAE
e-mail: martina_dickson@hotmail.com

M. Dickson et al. (eds.), *Gender in STEM Education in the Arab Gulf Countries*,
https://doi.org/10.1007/978-981-19-9135-6_9

student bodies in these fields. In some countries, the numbers of female graduates actually outnumber males in STEM and other programmes. However, further along the 'pipeline' as the famous analogy goes, the picture is different. In higher education, even in female-dominated subject areas aligned with so-called 'caring' professions like teaching, men still vastly outnumber women in top senior leadership roles. The reasons behind this are complex and include the existence of barriers to women progressing through academic ranks, which is highly dependent upon research output. Women still undertake by far the 'lion's share' of unpaid work—childcare, caring for elders, organising (if not carrying out) domestic chores, and the emotional work of child-rearing too. The vast majority of studies have indicated that this is often not true for men, and while men do face certain biases as fathers, they tend to spend more hours at the workplace and less hours performing unpaid labour in the home. They consequently often dedicate more time to their research. Men are also more likely than women to overestimate their own abilities and be more confident (Huang, 2013), are more likely to apply for promotions (which are necessary in the 'rise to the top'), and they are also more likely to benefit from known gendered perceptions in connection to brilliance, perceived by both males and females. The significance of the current study is that data presented on institutions' own websites is compiled in one publication, providing a 'snapshot' in time of the status of females in higher education and in STEM departments in the UAE. This allows for a clearer picture of gender balance (or imbalance), presenting a foundational framework from which to infer reasons for the status quo.

Literature Review

The Status of Female Leaders in Higher Education and Specifically in STEM Departments

Internationally, literature unanimously concludes that women are underrepresented in higher education at higher academic ranks and in senior management roles. This demarcation of roles in reporting gender representation is important, as research that presents only faculty numbers and/or departmental membership, misses this crucial information. In certain subject areas, particularly those traditionally perceived as being more 'female-friendly' such as the arts, education and social sciences, 'caring' professions such as nursing, and 'softer' sciences such as biology and healthcare, numbers of females are seen to be either equitable to those of males, or indeed to outnumber males. However, even in these subject areas, a closer examination of the data which takes into account the stratification per academic rank, and membership of females in senior management teams (including the decision-making positions at the top of institutions (such as Vice-Chancellors or Deputy Vice-Chancellors) indicates gender inequity issues in representation here too (Burkinshaw & White, 2017; Fitzgerald, 2013). Given that these positions have a critical power dynamic, it is concerning that women are far less likely to be found holding senior positions in higher education. It is also of concern that the lack of representation of females in senior roles means that female role modelling in these roles may not exist for younger female faculty members, and for students attending the institutions. This inequity is found to be true, regardless of the numbers of women graduating from those fields, and regardless of the numbers of female faculty. So, while numbers of female faculty are often presented as indicators of growing gender equity in higher education, the analysis per rank or position of this data frequently indicates that the highest positions are still extremely male dominated (O'Connor, 2019). This is unequivocally true in traditionally male-dominated fields such as STEM, but also (more unexpectedly perhaps) in traditionally female-dominated fields too.

This occurs for a number of documented reasons. Some of these are related to women being the bearers of children, and therefore the gender directly affected by the need to take maternity leave. Additionally, a huge body of research evidences that women do much of the childcare, particularly with young children. This includes the obvious, visible (yet unpaid) work of caring for children, such as washing, dressing, preparing food and perhaps less obvious work of a socio-emotional kind and the work of 'thinking', decision making and organization which precedes tasks. Women undertake far more unpaid work such as this than both their spouses and their male academic counterparts (Beagan et al., 2008; Bianchi et al., 2012; Fothergill & Feltey, 2003).

Schiebinger and Gilmartin (2010) write about ways in which female faculty holding senior positions in higher education are more likely to be effective in their careers if they 'outsource' their housework. This is not, they explain, equalising the division of labour between females and male counterparts, but it was found to be valuable in alleviating the load in some way, and providing some more hours in the day in which to focus on their careers. According to Schiebinger and Gilmartin (2010), when these women did employ others, this was mostly for core housework such as cleaning, and this correlated with higher levels of productivity (measured in numbers of publications).

Women are far less likely to be found holding senior positions in higher education. This is found to be true, regardless of the numbers of women graduating from those fields, and regardless of the numbers of female faculty. Sometimes numbers of female faculty members are presented as indicators of growing gender equity in higher education, however, the analysis per rank or position of this data quickly indicates that in the highest positions, importantly, the positions holding power and decision-making capacity, are still extremely male dominated (O'Connor, 2019). This is certainly true in traditionally male-dominated fields such as STEM, but also (more unexpectedly) in traditionally female-dominated fields such as teaching and nursing. Zhuge et al. (2011) attribute this global observation to be "due to deeply embedded unconscious gender-based assumptions that stall the advancement of women" (p. 637). In Wang et al.'s (2013) study of female leadership positions in China, only 7% of principal positions (Vice-Chancellor or equivalent roles), and just 10% of deputy level positions, were held by women. Even though over half of students, and close to half of faculty, in English universities are female, the vast majority of senior leadership positions are held by males (Shepherd, 2017).

As faculty attempt to progress along the academic career path and scale its ranks, at each stage in the promotion process, research output is at the centre of the successful transition to the next stage. In countries such as the USA, this process of 'tenure' is time-limited, and much has been written of the timing of tenure for female faculty, as it often coincides with child-bearing years. This leads to a pressured time of working against two clocks, one ticking towards tenure, and one ticking towards the diminishing of one's fertility (Acker & Armenti, 2004; Wolf-Wendel & Ward, 2015). Some universities have attempted to compensate for this pressure by modifying policies around tenure to include an additional year for parents. This has had mixed results, with some studies indicating that fathers used this time off, not to catch up on domestic tasks and become highly involved with childcare, but instead to have six months or a year off, free from the responsibilities of teaching and other time-expensive administrative duties (Antecol et al., 2018; Ward & Wolf-Wendel, 2004). This has led to a situation where men actually gained more of an upper hand with regards to readiness for promotion to the next rank. Other policies include lengthening of paid maternity leave (in some countries), and the option of paid paternity leave too. So, for some women, this is often a time to focus entirely on their home lives, while for men, it is often reported to be a time during which research continues or even expands.

Other contributory factors are at play too. Another reason suggested for having lower numbers of females in high academic ranks and positions of senior leadership is that fewer women than men apply for those positions in the first place. Women tend to underrate their own performance and abilities, while men tend to do the opposite, leading to a greater number of males than females applying for promotion opportunities (Baker, 2010). Men are more likely than their female counterparts to be associated with characteristics such as being brilliant (or a genius), whereas women's success is very often attributed to characteristics such as being hard-working, diligent and perseverant (Perez, 2019; Salin, 2020). Huang (2013) found that male academics held greater self-efficacy than females in computing, mathematics and social sciences. Another possible reason why women may be less likely to feel ready to apply for promotion at the same time as a man who has had the same career trajectory as them until this point, is that quite simply the woman may not have as many publications, and consequently does not apply for promotion due to not meeting minimal requirements. This is unlike males, who sometimes apply without the required number of publications (Baker, 2010). Women are much more likely than men to undertake fairly 'low-level' administrative positions, but these at least do contribute to one's promotional portfolio (albeit in a minor sense) as service (Shepherd, 2017). However much work goes on unrecognised too.

For women applying for promotion to higher academic ranks, where research publications become all-important in the 'publish or perish' climate of most academic institutions, some researchers have raised questions about gender equity in publishing itself, given that some studies clearly indicate lesser research productivity in women (Schucan Bird, 2011). Some research has suggested that gender bias is inherent in the use of publications, for example, that women operating within collaborative research groups are less likely to be named as first authors on publications regardless of contribution (West et al., 2013). Female authors have been found to be underrepresented in top journals within particular domains, and collaborative research projects which appear as publications tended to be from male academic teams (Teele & Thelen, 2017). Teele and Thelen also reflect that some of these top journals tend to favour specific research methodologies in their selection of articles to publish, such as favouring quantitative over qualitative research, which they see as also favouring male pursuits. Others have found that science research studies are more likely to be considered 'quality' if authored by males (Knobloch-Westerwick et al., 2013), and that females are less likely to publish research which relies upon 'expensive' funding (West et al., 2013). Research by Østby et al. (2013) contradicts this though and found that no significant gender bias in either citations or publications was found. Whichever is true, perhaps the latter leads into a situation which Reed et al. (2011) describe, whereby women are found to be in fewer positions of leadership than males, even where publication rates equal or even exceed those of males towards later career stages, and similarly, females held fewer positions of the highest academic rank (e.g., full professors) than males over their career journeys. Reed et al. (2011) described how this "paucity of qualified women in leadership positions both deprives organizations of the unique skills and perspectives women bring to such roles" (p. 46).

The helpful 'leaky pipeline' analogy is used by Van den Brink and Benschop (2012) to chart the transitions at each stage of the pipeline—from recruitment of new job starts, to appointments at first promotional stages. Some key findings from this study are that (1) they found that similar percentages of females and males who had applied for senior leadership positions were also interviewed (indicating that bias interfering with selection for the interview was not the reason for the 'leak'), and (2) that in some fields, the number of female applicants to senior leadership positions were disproportionately low when compared to the number of qualified and eligible female academics, suggesting that more subtle, nuanced barriers are in effect. They discussed possible gender equality practices that are specifically designed to undo dominant equality practices, which included proactively taking steps to recruit women, appointing women as chairs with a special responsibility for women's issues, and making policies and procedures for mentoring and recruitment fully transparent (Van den Brink & Benschop, 2012).

One argument that is often put forth by way of explanation as to why females are less likely to rise in academic ranks, or take up senior leadership positions, is that females in academia may be more likely to focus on teaching and pastoral responsibilities, perhaps taking up positions that align with this but which do not particularly contribute towards a promotional system that relies heavily upon research productivity and publication (Winslow, 2010). Having these preferences (if indeed, they are preferences), and the associated responsibilities means that a greater proportion of professional time is allocated to this type of work, and consequently, less to the all-important research portion of one's academic work. Winslow argues that the gender inequality in academia is "produced and maintained in this aspect of employment"

(p. 769), and that the consequent gaps in inequitable timing allocations between research and teaching time are not explainable purely by "preferences or educational and institutional attributes" (p. 769). Morley (2013) writes of the interaction of power and gender in higher education leadership, and ways in which concepts of academic meritocracy sometimes mask nuances beneath a surface rationality, that require recognition of female leadership capacity in a way that is both accurate and nuanced. Similarly, acknowledgement of what many researchers have indicated—that there is a profound gendering in academic activity—is likely to aid understanding of why, and how, the underrepresentation of females in senior roles, occurs (Savigny, 2014).

The omnipresence of the issue has inspired much research. Airini et al. (2011) were guided in the research to find out about what "helps or hinders women to advance in university leadership roles, as reported by women" (p. 48). They found that women benefited greatly from meaningful professional development that included working with a mentor to increase self-awareness in areas such as opportunity-seeking and taking, and 'soft skills' such as improving time management. This enabled the women in the study to prioritise factors that would have most impact on leadership advancement, including applying for and successfully securing, research grants. Other key help or hindrance factors that arose from this study were the environment of work itself, relationships at work and at home, and what Airini terms 'invisible rules', which allude to the same masks of nuances to which Morley (2013, 2014) alluded.

Some interesting theoretical frameworks are put forth by Heijstra et al. (2017) with regards to academic housekeeping, and the way in which much of academic housekeeping in universities does indeed tend to be carried out by females. By 'academic housekeeping', Heijstra et al. mean the administration work typically involved in middle management positions such as programme chairs, departmental heads, line management approvals, faculty evaluations and appraisals, reporting of indicator data to their institutions and other such administrative duties. The time associated with these types of positions are often colossal, yet the work involved does not typically contribute significantly to promotional pathways for faculty. Heijstra et al. (2017) explore another element to the performance of academic work, arguing that there is sometimes the unequal distribution of academic housework between newcomer and more established faculty. Fitzgerald (2013) writes poignantly and personally of the loneliness and isolation in leadership roles, attributed by Fitzgerald to be at least in part due to the "marked absence of women in mid-senior positions in the places" (p. 4) in which she found herself.

Work by MacFarlane and Burg (2019) showed that female professors are more likely than males to conceptualise their professorial roles as having strong academic citizenship work, such as mentoring, compared to their male counterparts. They described this as "an obligation that weighs especially heavily on women working in science, technology, engineering and mathematics areas" (p. 262). They see that, while this indicates how 'academic housework' can continue to create barriers for women's academic careers, it also illustrates a strong commitment to an intellectual leadership role. Far from needing to be corrected, perhaps this is an indication that male professors should reflect on their own roles and adapt in this way. Both males and females, in this study by MacFarlane and Burg (2019), had similar understandings of the research leadership aspect of their roles.

Airini et al. (2011) describe female academic participants as having made conscious choices to prioritise time with their families, with full awareness that this may come at a cost or expense to their research, and therefore to their career development. One of the participants in the Airini et al. (2011) study, for example, acknowledges her slower progressional career pathway, but states that:

> for me personally this was a choice. Having family is a big part of my life and I do not work evenings or weekends … I have no doubt this results in lower research productivity, but so be it (p. 54).

Airini interprets this as a choice which was both a satisfying and personal choice, which suggests agency, albeit with career consequences. This idea, that a woman's choices are a major contributing factor to low uptake of leadership roles, and low numbers of females at high ranks, is one school of thought. However, Shepherd (2017) takes a different view, and argues that such ideas—which suggest that women have agency to make these choices, and that these choices are responsible for the resultant gender inequities at the end of the rhetorical pipeline, do not take into consideration structural inequities which often exist inherently in institutions. In addition to this, Shepherd (2017) also challenges the "notion of women's missing agency—characterized by a lack of confidence or ambition and a tendency to opt out of applying for the top jobs—as an explanation for their continued underrepresentation" (p. 82). STEM fields in particular suffer from an 'image' problem too, with the associations of STEM as being a career with long working hours, and "relentless focus", one which requires the "giving up [of] family life … due to the belief that household and family work are women's responsibilities" (Canetto et al., 2017, p. 4).

Where males continue to dominate in leadership roles, particularly those at the very top, the situation may perpetuate. This is also exacerbated by the fact that organizational heads themselves may be involved in the selection of their vice-positions, and may select people who reflect their own demographic, leading to less diversity. Then of course, because someone with more experience would be considered less risky in recruitment procedures for leadership roles, and therefore more likely to be selected for this and other similar roles. So, it is easy to see how perpetual gender imbalance, or any imbalance routed in power differentials, becomes cyclic.

Higher Education in the UAE

In the UAE, the higher education sector has grown from having only five higher education institutions in 1990 (Sim, 2020), to over eighty at the time of writing. The dramatic expansion of infrastructure in the country, including the education sector, begin in earnest in the early 1960s not long after the discovery of oil, which led to a change in national wealth almost overnight. In addition to the federal universities in the country, there are a large number of private institutions that cater to both national and expatriate students. Some of these are also international ‘satellite’ campuses of well-established international universities. As a result of this diversity, the diversity of faculty working in the institutions is also considerable. A large proportion of faculty are expatriates from non-Gulf Arab countries, a much smaller proportion are Emirati nationals, and the others are expatriates from other countries. The relatively small proportion of national faculty in higher education is an issue which some recruitment policies have been put into place to overcome, such as Taqweem, the national agency for the recruitment of nationals, who act as a central employment application centre for many sectors. There are also policies in place to attempt to increase the numbers of nationals in workplaces, such as sometimes offering positions with lower qualification requirements (e.g., a master’s instead of a doctoral degree) in order to attract more nationals. This is an issue which exists for various reasons. Firstly, the relatively recent development of graduate programmes in the UAE has meant that in earlier decades, those who wished to pursue graduate degrees would often do so overseas, and due to cultural reasons, this was more likely to be males than females. It is also the case that highly qualified nationals, particularly those in STEM fields, are often more likely to take up employment in industrial STEM fields, perhaps due to financial reasons.

Both females and males in the UAE have wide and supported access to higher education. This is free for nationals in federal institutions, and often accommodation and transport are offered free to undergraduate students as well. The majority of students in government higher education institutions are female, and females outnumber males in all majors apart from engineering (Findlow, 2007). Some researchers have argued that, while the rates of attendance and graduation for women in these institutions has increased dramatically, “for Emirati women, education is not always a means to an end, but simply the end itself. For some women, education is related to supplementing personal growth and contentment” (Sim, 2020, p. 73). This is sometimes explained in terms of the increased social status that higher education confers upon both females, males and their families (Engin & McKeown, 2017). In other words, higher education studies do not necessarily link to the choice of career. This was evidenced in a study by Dickson and Tennant (2019) which showed that for some women, the completion of their degree was enough; they had no plans to join the workforce. That study took place in the field of teacher education, which may appeal to more conservative families, so the findings may be less applicable to STEM fields.

The UAE also has some of the highest rates of female student registry in higher education in the world (Sims, 2020). Both of these facts might lead one to hypothesise that a greater proportion of female representation at both senior academic ranks, and positions of senior leadership, would be expected. Although women are educated to a high level in the UAE (Madsen & Cook, 2010) there is also a ‘leaking pipeline’ of women from education into senior managerial and organisational roles (Afiouni & Karam, 2017). The governing accreditation body in the UAE—the CAA (Commission for Accreditation Agency), from where the list of institutions for this study was procured, themselves have just one female commissioner among their team of twelve, perhaps sending a questionable message to the institutions they review and accredit in terms of prioritising gender equity.

The UAE National Vision 2021 highly emphasises the participation of both female and male national citizens, to enrol in higher education and ultimately to participate in the national workforce. This workforce, it is hoped, will be skilled enough to contribute to the growth of the future UAE Knowledge Economy. The UAE government, and the legacy laid forth by Shaikh Zayed Al Nahyan, ensures that Emirati women have been given many opportunities within both policy and practice to succeed in their education and in the workplace; indeed, there are many examples observed frequently in public media of Emirati women who have excelled in STEM careers in the UAE (Sim, 2020). The UAE is, however, a country with a majority of expatriates within its resident population, and also a large number of expatriates employed as faculty in higher education, adding further complexity to the gender equity discussion.

In the UAE, some of the constraints noted in literature are alleviated by the propensity of ubiquitous and inexpensive domestic help, which may have a positive impact on the time pressures and well-being of academic mothers (Dickson, 2019). In theory—because women in the UAE have access to support which other women do not (inexpensive domestic help)—one would expect greater representation at higher levels of leadership, primarily because the literature strongly suggests this would be the case. If we do find evidence of this here in the UAE, this is empirical evidence in support of the theory that a key reason for gender inequity in representation is indeed the 'second shift' performed by women at home. If not though—i.e. if we saw no real difference in equity patterns in the available data—from this we can make theoretical projections as to why not, because this indicates that the reason women are not reaching these levels is much more complex and goes beyond the domestic challenges cited by other literature.

The impact of this work is accentuated by having one publication giving a detailed picture of gender representation in higher education. There is no such study, currently, in the UAE, and as such, this paper will potentially hold interest for both higher education administration and policy makers. In the discussion and implications section of the chapter, we discuss the status quo and put forth suggestions and recommendations on the basis of our findings.

This research aims to quantify, using institutional websites with faculty profile information, the gender profiles of faculty per rank and senior management representation, in order to assess gender representation in these positions. A study of the current situation in higher education, therefore, is not only prudent, but crucial, in assessing the status quo and by doing so, providing a framework upon which to build policy and practice which is gender equitable and inclusive.

The following research questions are explored through this study:

RQ 1. How do proportions of gender of faculty across all HE departments compare to those in STEM departments?

RQ 2. What relative proportions of both STEM and non-STEM faculty members in higher education institutions in the UAE are males versus females, and what are these proportions at each academic rank (lecturer, Assistant Lecturer, Lecturer, Assistant Professor, Associate Professor, Professor)?

Methodology

The data sources that will be used in this study are publicly accesible institutional websites. Included in the website analysis will be all higher education institutions in the UAE that are accredited by the national accreditation body, the Commission for Academic Accreditation (CAA). These institutions are outlined on the CAA's website.[1] We recorded the following data per institution systematically:

- Name of institution and departments (or colleges, as appropriate to the particular institution)
- Total number of faculty, and total numbers of male and female faculty members across departments, and specifically in STEM departments
- Total numbers of males and females at each of the published academic ranks (namely, Assistant Lecturer, Lecturer, Assistant Professor, Associate Professor, Professor) in all departments, and in STEM departments
- Gender of university heads (DVCs or equivalent role, VC or equivalent role) and other senior leadership posts
- Academic rank of those working in senior leadership roles
- STEM departmental leadership: gender and academic ranking

From these, descriptive statistics were generated about gender representation in each of these institutions, and overall across the UAE. For most universities in the UAE, this information is readily available on their websites (see examples[2]).

There were some limitations to this approach, outlined subsequently. Where websites do profile individual faculty members with all of the required data except photographs, names were cross-checked among the researching team (PI and Research Assistants) to attempt to moderate potential name-gender errors. In the case of uncertainty, the particular faculty name was cross-referenced with publications (e.g. through Google Scholar) and through professional networking sites. Where websites did not profile individual faculty members, the institution was excluded from the study. Where websites do profile faculty members (e.g. departments, biographies and photographs) but do not include information regarding academic ranking, these are included in the data for reporting for faculty gender only, and not academic rank.

Finally, any website at any time can only be considered a 'snapshot' of the institution at that point. Even if newly hired faculty members are not yet represented on the website, the same could be argued of older faculty who are still there and not yet

1 https://www.caa.ae/pages/institutes/all.aspx.

2 https://cos.uaeu.ac.ae/en/departments/biology/faculty.shtml (UAE University).
https://www.adu.ac.ae/study/colleges/college-of-arts-and-sciences/Faculty/ (Abu Dhabi University).

removed. However, it is a depiction of the representation at that point, and unless the site has not been updated in the past five years or so (highly unlikely, and this would be evident in any case from hallmarks of the site in other ways, such as faculty publications), then we did not consider this to be a major methodological issue. The site analysis took place in the middle of an academic year, as we reasoned that this would be far less likely to be problematic than at the beginning of an academic year.

In terms of risks to either individuals identified through this website analysis, these were eliminated by reporting data within departments across universities. For example, data is be reported as "80% of faculty members in science departments across the country are male", rather than "In X University, 8 out of 10 science department members are males". This sensitivity is borne dearly by us as researchers, even although since this information is already publicly available on the university websites. We also report aggregated data from universities as a whole, e.g. X% of faculty across UAE universities are males, and of these, X% are appointed at the rank of Associate Professor or Professor. Data regarding senior leadership roles, since there are far fewer of these, was handled sensitively and as aggregated data only.

Findings

RQ 1. How do proportions of gender of faculty across all HE departments compare to those in STEM departments?

If we look at the total number of faculty (differentiated by gender) in HEIs across all departments, compared with the same faculty in STEM departments, we see that very similar percentages occur in both STEM and non-STEM departments. There are slightly less females in STEM than in all departments, and slightly more males in STEM than in all departments, but these percentage differences are very small (Table 9.1).

RQ2. What relative proportions of both STEM and non-STEM faculty members in higher education institutions in the UAE are males versus females, and what are these proportions at each academic rank (lecturer, Assistant Lecturer, Lecturer, Assistant Professor, Associate Professor, Professor)?

Comparing academic ranks of faculty across departments shows us that the vast majority of faculty employed at professorial role are males (10.5% versus 27.8%). See

Table 9.1 Faculty in STEM and non-STEM departments by gender

Department	Females	Males
All departments	1988 (32% of all faculty)	4295 (68% of all faculty)
STEM departments (e.g. physics, chemistry, biology, engineering, computing/technology)	959 (29% of STEM faculty)	2384 (71% of STEM faculty)

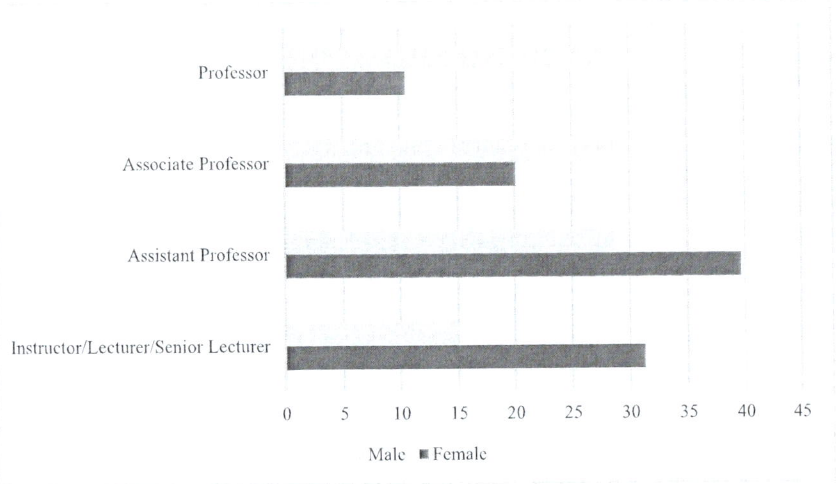

Fig. 9.1 Comparison of percentages of academic rank per faculty gender (all departments)

Fig. 9.1 for a pictorial representation of this data. At the rank of associate professor, the percentage of males are still higher than females, but the ratios are closer (20.1% v 28.6%). However, at the lower ranks of assistant professor, a higher percentage of females than males is seen. Then, at the lowest ranks (lecturer, instructor, and equivalent), the percentages of female faculty are more than double that of males.

If we look at the percentages of faculty working as professors in STEM fields as a percentage of the whole STEM faculty, as opposed to as a percentage of their own gender, we see another pattern. Table 9.2 shows that 58 female and 543 male professors are recorded in STEM departments, a massive gender imbalance. So, females make up almost 10% of all professors in STEM departments, and the equivalent figure for professors in non-STEM departments is 14.5%. Since we also noted that only 8% of those holding the most senior roles in institutions are females, with the vast majority likely to be professors, an overall 8% is not dissimilar to the numbers of professors (10%) across faculty, not in leadership roles. In other words—the figures of female representation in leadership (if a professorial ranking is taken to be the most common rank) actually appear to broadly reflect the percentage of female faculty in the country too.

Figure 9.2 shows graphically that there is a substantial difference in the percentages of females and males at each academic rank in STEM departments, looking at the faculty make-up as a whole. Note that all of the percentages add up to 100, since Fig. 9.2 shows the percentages of academic ranks for faculty overall. Key features to note are, for example, that of the total number of faculty in STEM departments, only 2% are female professors, and only 4.9% are female associate professors (compared with 21.7% male associate professors). The majority of female faculty are assigned ranks of assistant professor and below, and the percentage of faculty in the ranks of lecturer or senior lecturer are roughly gender balanced.

Table 9.2 Faculty in STEM and all departments by gender and academic rank

Department		Academic ranking							
		Professor		Associate professor		Assistant professor		Lecturer, instructor or equivalent	
STEM (e.g. physics, chemistry, biology, engineering, computing/technology)	Total number	Female	Male	Female	Male	Female	Male	Female	Male
		58	543	141	629	225	460	391	225
	% per gender	6.6	26.8	16.8	31.1	25.7	22.7	51.5	19.3
	% per total faculty	2	18.7	4.9	21.7	7.8	15.9	15.5	13.5
All departments (e.g. languages, arts, social studies)	Total number	Female	Male	Female	Male	Female	Male	Female	Male
		187	1094	328	1126	703	1125	555	587
	% per gender	10.5	27.8	20.1	28.6	39.7	28.6	31.3	14.9
	% per total faculty	3.3	19.2	5.7	19.7	12.3	19.7	9.7	10.3

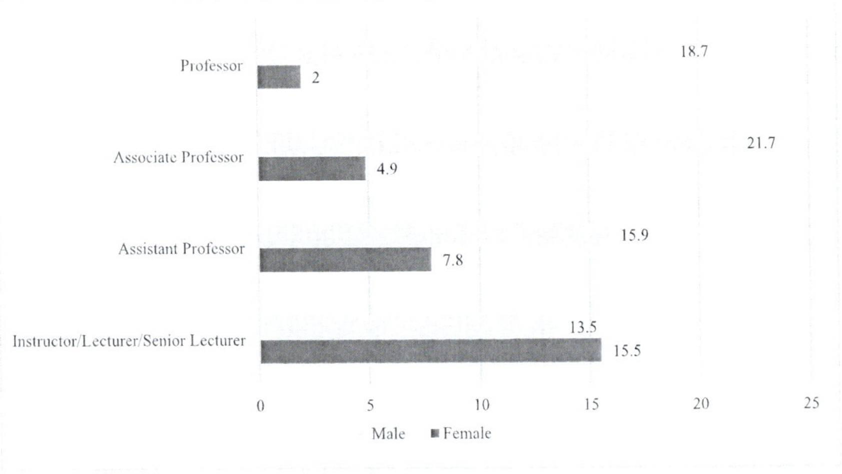

Fig. 9.2 Comparison of total percentages of male and female faculty, per academic rank (STEM departments)

An important statistic is that of the percentages of academic ranks within one's own gender, which has important implications for notions of social norms and role models. Of all female faculty in STEM departments, only 6.6% are professors, compared with the percentage of male faculty in STEM departments who are professors (26.8%). The percentages of associate professors who are female in STEM departments is half that of males (Fig. 9.3). Percentages of one's own gender are closer for assistant professorial ranks (25.7% female versus 22.7% male), but a dramatic change is seen when we analyse the data of those at the lowest academic ranking, that of the instructor, senior lecturer or lecturer. Here, we see that more than half of all female faculty working in STEM departments are employed at these ranks, compared to only a fifth (19.3%) of males.

Finally, Table 9.3 shows the breakdown of departmental leadership in STEM and non-STEM departments. We can see that there is a higher percentage of female departmental leadership in all departments (29%) compared to in STEM departments (15%). This approximately echoes the pattern of percentages of female faculty in all departments (14.6%) compared to in STEM departments (9.6%).

Discussion and Implications

Comparing academic ranks of faculty across departments shows us that the vast majority of faculty employed at professorial role are males, and with decreasing academic rank, the ratio of females to males increases. It is concerning, particularly in a country where the majority of students in higher education are female, that at

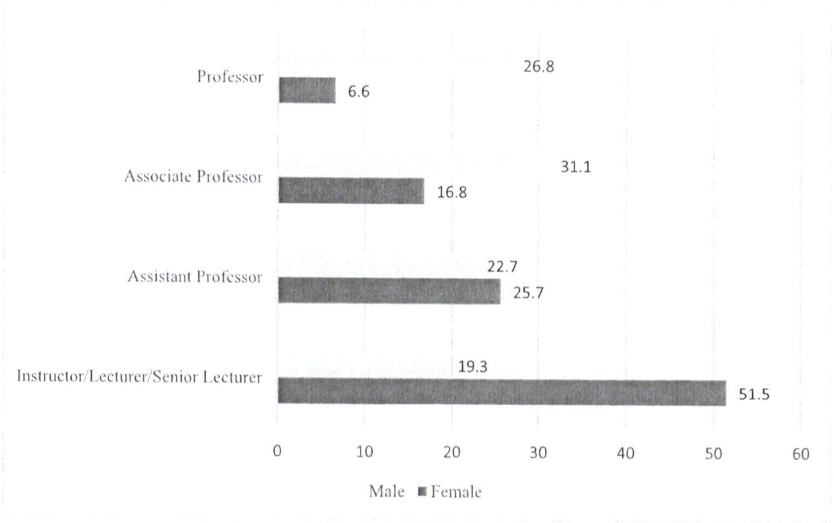

Fig. 9.3 Comparison of percentages of academic rank per faculty gender (STEM departments)

Table 9.3 Department leadership in STEM and non-STEM departments, by gender and academic rank

Departments	Gender of department head	
STEM (e.g. physics, chemistry, biology, engineering, computing/technology)	Female	Male
	23	150
All departments	Female	Male
	44	152

the lowest academic ranks (lecturer, instructor, and equivalent), the percentages of female faculty are more than double male faculty. So, there is gender imbalance on multiple levels. Firstly, in STEM fields, we see that the percentage of female faculty is much less than males, and (predictably), the percentages of female faculty who have reached the professorial level is only 6.6% in STEM departments, compared with the percentage of male faculty who are professors (26.8%). But the problem persists outside of STEM departments too, which is different from in some other countries, such as in the UK, where around half of the faculty in higher education are female (Macfarlane & Burg, 2019). As such, at the beginning of this higher education 'pipeline', there is already a stark inequity in gender representation. Moving further up the academic ranking, we see that the percentage of associate professors who are female in STEM departments is half that of males, but the percentages of one own's gender at the rank of assistant professor, are similar. The implication of this is that the 'playing field' from which faculty work and apply for promotions and leadership positions is simply not level. The problem is not *only* that of gender imbalance in leadership positions in STEM, and not *only* the relatively low numbers of females in both STEM and non-STEM departments (although these are both major challenges). The key issue to reflect upon is also the striking, disproportionately low academic ranking of females in both STEM and non-STEM fields.

While we see from the data that there is a higher percentage of female departmental leadership in all departments (29%) compared to in STEM departments (15%), this in itself is not particularly surprising, as it approximately echoes the pattern of percentages of female faculty in all departments (14.6%) compared to in STEM departments (9.6%). Within the literature on female STEM representation, there are various schools of thought by way of explanation for why there are so few females in STEM leadership positions, and indeed in academic leadership in general. One of the theories around this is that women face barriers during the application and interview phases, while others posit that it is due to the lack of support experienced when actually in these roles, leading to attrition. Zhuge et al. (2011) lean towards theories of gender-based assumptions that likely stall achievement, which may be relevant here. Other themes suggest that the imbalance arises due to males having greater self-efficacy (Huang, 2013), leading to them applying more frequently for promotion. Morley (2013) also talks about gender inequality as being characterised as a critical deficiency in leadership in higher education.

The stark fact, highlighted by the data presented in this study is that policies must first focus on supporting the increase of the numbers of females at higher academic ranks, since the vast majority of female academics are not meeting the criteria to even apply for these ranks or positions in the first place. Why are more than half of the female faculty in science departments employed at the lowest academic rank (the entry ranks, essentially), of instructor and lecturer? While we see clearly that there are issues around the numbers of females in raw figures (only 29% of faculty in STEM departments are females), and this 'supply' of female faculty from the university needs to be addressed (more on this later), this does not explain the dramatically differing percentages of females working at the ranks of associate professor and professor (23.4%), compared to those working at the rank of assistant professor and below (76.6%). Given that most universities have policies or practices around the need for faculty to be established at these academic ranks in order to be eligible for applying for more senior leadership roles, either at a departmental level, or beyond this into the top tiers of institutional management, these low figures of eligible females are deeply concerning. The status of female professors in the UAE is more dire than in other countries, such as in the US, where the gender disparity between professors was stated in the Digest of Educational Statistics (2015) to be 2.3:1.

The fact that the majority of females are at a rank of below or at assistant professor implies a number of things. One, that they are not eligible to be employed at these ranks upon initial entry to their institutions, and two, that they are not able to meet the requirements of promotion once within the system. For most universities, promotion between academic ranks requires evidence of competencies in three key areas: research, teaching and community service. The latter is usually attributed a lower weighting than teaching and research. In the teaching category, typical performance indicators would be aspects such as: line manager evaluations, teaching philosophies, samples of teaching plans, quality of lesson resources and student evaluations. In the research category, the performance indicators tend to be much 'harder' quantifiable figures, such as: numbers of publications in ranked journals, numbers of citations, awards of grant funding, collaborative research partnerships and in universities with a more commercialised element, or emphasis on industrial partnerships, perhaps elements such as patents would contribute too. One of the problems with this model is that these tend to be more difficult for junior faculty to achieve in the first place. Competitive grant funds, for example, are more likely to be awarded to academics with higher academic rank, since this signifies to funding bodies experience and perhaps a greater likelihood of successful grant management. The highest academic rank of professor (sometimes called 'full professor') has a large element of research leadership in the promotion criteria—faculty would generally be expected to show that they had the experience of leading research teams, handling sometimes large financial budgets and overseeing project management.

These are difficult elements for faculty at lower academic ranks to gain experience in. Much has been written about the gendered nature of research and of publication prolificity. Literature suggests that for academics, the early to mid-career phase (when one is experienced enough to begin to write prolifically and perhaps to be in a situation to apply for grant funds), often corresponds to a time where having children would be most likely (resulting in analogies around ticking biological clocks). The incumbent time away from one's career, or the reduced focus on one's career through either maternity leave, or simply the mental exhaustion of raising young children and all that this entails, may simply mean on a practical level that research—the activity most likely to lead to promotion—is placed on a back burner and secondary to more 'urgent' elements such as teaching responsibilities. This cycle places the woman at a disadvantage, since, as time goes on, they are still not meeting the requirements for promotion to the next academic rank. As mentioned earlier, there is research evidence to show that males and females understand their professorial roles differently. While both genders indicate an understanding of the importance of research leadership as a prime factor in the role, women were much more likely to emphasise elements of their academic citizenship, such as mentoring. Savigny (2014) writes of the inherent gendering of academic activity, which is a contributory factor to these low percentages of female academics working at high ranks. For those women who do make it to the top of STEM academic roles, leadership can also be extremely lonely. Fitzgerald (2013) describes this loneliness and isolation as feeling like an 'outsider', which she attributes in part to the lack of other women in those positions. So, in this sense, it becomes something of a vicious cycle, as other women see these positions as less appealing, and so are less likely to aspire to them, and so on.

15% of leaders of STEM departments are females, but this means that 23 out of 959 female faculty in STEM departments (just over 2%) have departmental leadership duties. It is difficult to tell whether this corresponds to academic rank, since this data was not available for all institutional websites. However, if we look at the number of male departmental leaders in STEM fields, which is 150/2384 (6%), this is triple the number of females. But there are just over double the numbers of males, compared to female STEM faculty, indicating that the proportion of faculty who have taken on STEM departmental leadership roles is more than the ratio of females to males. Placing this in the context of the fact that the vast majority (over 76%) of female faculty in STEM departments have a ranking of Assistant Professor or below, this suggests that there is a disproportionate amount of departmental leadership led by females. The other important point for consideration is that departmental roles are often seen to be not dependent upon academic rank. Where leadership of a department (or college, sometimes) was undertaken by a role named 'Dean', these were mainly holders of professorial roles. However, when the title was Head, Chair, Coordinator, or other equivalent titles, there was no correlation to rank. Of the STEM departmental data available to us, more than 50% of faculty in those roles were Assistant Professors and even lecturers and instructors (in smaller institutions) are performing these roles. This interesting scenario indicates that, where the rank of faculty is not required to perform leadership on a small-scale (such as departments), female representation is seen more equitably than in the highest leadership position roles. This has important implications, as we shall see subsequently.

The term 'academic housekeeping', coined by Heijstra et al. (2015), covers the logistical, administrative, holistic, service work that is often essential to the effective running of academic institutions. Heijstra et al. (2017) argue that academic housekeeping affects academic capital and increases inequities. Kalm (2019) describes these tasks as being 'invisible', time-consuming, low status, but necessary, yet the sheer work involved in this type of task inhibits or delays, promotion 'readiness' and likelihood (Misra et al., 2011). No doubt, early career males can also experience this effect, however the roles of the 'academic housekeeper' are more often undertaken by women for two reasons, either that women are more likely to volunteer for those roles, or, that they are more likely to be selected or groomed towards those roles. This could occur for a number of different reasons, mostly housed within notions of females as being more caring, empathetic, better organised, with more of an 'eye for detail' (Guarino & Borden, 2017), particularly where the role involves interactions with students. While these are important roles—certainly some female faculty take these on with relish, completely voluntarily and with the rewards of job satisfaction which this type of work can bring—they also take up copious amounts of time. Time is required to perform all of these administrative tasks, some of which are often fairly low level intellectually, and in order to align with institutional quality assurance demands. There are related meetings, paperwork, unplanned impromptu drop ins by students and faculty, and generally being 'on call' to trouble shoot and 'fire fight'.

The consuming nature of these roles makes it less likely that the faculty member serving in this way will be able to carve out vital time for the work of research—writing research proposals, applying for grant funds, collecting and analyzing data and then writing publications for research dissemination. The need for quiet or at least uninterrupted space in which to perform much of the thinking, writing and creative thought work around research is critical and often difficult for women to find. This has become even more problematic during the recent Covid pandemic, when many academics have been working from home and simultaneously supervising children's online learning, to compound matters (Miller, 2021). What may seem like a good opportunity to an aspiring academic professor, as an early career researcher, is very likely to become detrimental to faculty research.

So—what is the solution to this? How much of a responsibility do academic leaders have to ensure that junior faculty do not overload themselves with administrative work or academic housekeeping to the point that they effectively take themselves out of the running for future rank promotion opportunities? Should these positions simply have the requirement of being associate professor or above? Or, should people have the agency to make these decisions by themselves?

The other main question to consider is this. We have established that the numbers of 'eligible' females for senior leadership positions are currently so low that simply creating policies to encourage more applications from females, or to retain females in these positions, would be insufficient to even begin to address the dramatic gender inequity issues. The low percentage of female faculty in STEM departments in HE is of great concern and in need of address, but the numbers of female faculty at low academic ranks are also of great concern. Heijstra et al. (2015) discuss the often-slower pace with which women climb the 'academic ladder' due, primarily, to family-related variables, but this is seen to be true even where family-related variables are not negatively impacting the woman's journey. We also have no way of knowing if all of the women have parental or other caring responsibilities. Among the many consequences to an institution, to other faculty, and to students, there are dire consequences of persistently remaining in low academic ranks for women too. In the UAE, unlike in other countries such as the US, for example, there is no particular requirement to raise oneself through academic ranks within a specific timeframe, or indeed at all, in order to retain employment. Consequently, women experience lower salaries, less decision-making agency, and possibly, in some institutions, higher teaching loads.

The other poignant question is why more females are not joining higher education as academics. This is not simply a 'pipeline' problem in the UAE, since many STEM degrees have near equitable numbers of females and males. This is a general global pattern, but it is actually pronounced in the Gulf countries, possibly related to the fact that secondary school girls repeatedly outperform their male counterparts in standardised assessments such as PISA (Melkonian & Areepattamannil, 2018). In Saudi Arabia, the number of females are currently outnumbering males in some science degrees, for example. In Khalifa University, here in the UAE, the number of female applicants to science degrees outstrips those of male applicants, and their graduates are near gender equitable. In the UAE as in other gulf countries, there is a high dependency upon expatriate skilled expertise, and in fact these make up the majority of the faculty in higher education institutes. So clearly—a more proactive and inclusive approach to actively recruiting female academics in STEM fields is necessary, including by mechanisms such as 'head-hunting' via professional networking sites. There needs to be deep thought and reflection as to how to attract graduating females into academia, in addition to the traditional routes of industry and corporate STEM fields.

The other key element discussed in literature is the amount of domestic work that women do, in comparison to males. We stated in the literature review that, according to Schiebinger and Gilmartin (2010), female academics in higher education were more effective in their careers, and more likely to be research-productive if they 'outsource' their housework. But in the UAE, this pressure is alleviated for most homes by the ubiquitous employment of domestic help. A recent study of academics in the UAE indicated that more than 90% of them employed domestic help for housework, chores and perhaps childcare (Dickson et al., 2023). This is not to say that paying someone to help with domestic chores completely frees one from responsibilities in that arena, but that it is perhaps less likely to exacerbate responsibilities. One thing that it does not absolve either parent from, and we know that this is much more likely to fall to the female, is the 'thinking work' of parenting. That is, the emotional work of parenting, which is again more likely to fall on females (Azar et al., 2008), and persists whether one pays someone to carry out chores or not. Perhaps this is one of the reasons that we do not see a correlational raise in female representation even where women have done as Schiebinger and Gilmartin (2010) suggest, and 'outsource' their housework.

The lack of female representation creates issues of a lack of visibility of women for young women too, particularly students who might potentially be attracted into academia in the future. In Khalifa University, for example, which is the top STEM university in the UAE, there are male heads of departments in the vast majority of STEM departments, and the small number of female heads of departments are only in the biomedical department. In another large UAE institution, Fatima College for Health Sciences, where the vast majority of both students and faculty are female, the departments are led by males (according to the institutional website). What is the potential impact of this, both systematically and symbolically, on the females in the institution? Canetto et al. (2017) posit that women are constrained by limiting practices and gender ideologies that affect their likelihood of taking up faculty positions. Perhaps these elements in the UAE too, are affecting the attractiveness of academia as a career.

Shepherd (2017) describes the situation for women in academia who are both ambitious and highly competent, but who may be being disadvantaged by structural issues. Shepherd mostly indicated that these factors included elements such as selection and recruitment processes, but as we have highlighted in this study, these factors are not relevant if the females themselves are not eligible for the recruitment in the first place due to low academic ranking. We do agree, however, that the reasons behind this (the appointment of females with low academic ranks to academic housekeeping roles, among others) are also structural. This implies that, exactly as Shepherd says: "women's missing agency is not in itself an adequate explanation for their continued underrepresentation at the top of higher education" (p. 82).

Fitzgerald (2013) argued that female leadership should be made more visible, and if so, then opportunities may arise for women's voices to be listened to. The implication of this is that other women might be inspired by this, but this inspiration will be meaningless (surely) if they cannot raise their academic ranks in the first place. We suggest that a greater emphasis should be placed upon rising through academic ranks as a primary goal, as opposed to encouraging women to apply for leadership posts as a sole emphasis. The current research would also be complemented by further research of a qualitative nature, using follow up interviews with female STEM faculty to truly understand the reasons behind the relatively low numbers of females in higher academic ranks and positions of power, including gaining an understanding of any structural barriers that face females in the field.

Conclusion

Our study has highlighted several key issues facing females in academia in the UAE, both in STEM, and in other departments too. Female faculty are underrepresented across higher education, and grossly underrepresented in STEM departments. Perhaps consequently, the numbers of females in senior leadership positions is small in comparison with males in similar positions, yet there are disproportionate ratios of females in departmental leadership or coordination positions where academic rank is not a prerequisite. The number of female academics in the ranks of associate or full professor are dwarfed in comparison to the numbers of their male counterparts. Involvement in the 'academic housekeeping' work of lower-level administrative roles has a profoundly negative impact on the time that a female academic can spend on her research, yet ultimately, this is what decides whether or not she can apply for promotion to higher ranks. Institutions in the UAE must re-focus their efforts on redressing this imbalance by two main pathways; firstly, by encouraging and supporting females to raise their academic rank and not to take on 'academic housekeeping' roles until the rank of associate professor is met, and by adopting policies and practices that would encourage greater numbers of female faculty into higher education.

References

Acker, S., & Armenti, C. (2004). Sleepless in academia. *Gender and Education, 16*(1), 3–24. https://doi.org/10.1080/0954025032000170309

Afiouni, F., & Karam, C. (2017). Debunking myths surrounding women's careers in the Arab region: A critical reflexive approach. In *Career guidance and livelihood planning across the Mediterranean* (pp. 55–70). Brill Sense.

Airini, Collings, S., Conner, L., McPherson, K., Midson, B., & Wilson, C. (2011). Learning to be leaders in higher education: What helps or hinders women's advancement as leaders in universities. *Educational Management Administration & Leadership, 39*(1), 44–62. https://doi.org/10.1177/1741143210383896.

Antecol, H., Bedard, K., & Stearns, J. (2018). Equal but inequitable: Who benefits from gender-neutral tenure clock stopping policies? *American Economic Review, 108*(9), 2420–2441. https://doi.org/10.1257/aer.20160613.

Azar, S. T., Reitz, E. B., & Goslin, M. C. (2008). Mothering: Thinking is part of the job description: Application of cognitive views to understanding maladaptive parenting and doing intervention and prevention work. *Journal of Applied Developmental Psychology, 29*(4), 295–304. https://doi.org/10.1016/j.appdev.2008.04.009

Baker M. (2010). Career confidence and gendered expectations of academic promotion. *Journal of Sociology, 46*(3), 317–334. https://doi.org/10.1177/1440783310371402.

Beagan, B., Chapman, G. E., D'Sylva, A., & Bassett, B. R. (2008). 'It's just easier for me to do it': Rationalizing the family division of foodwork. *Sociology, 42*(4), 653–671. https://doi.org/10.1177/0038038508091621

Bianchi, S. M., Sayer, L. C., Milkie, M. A., & Robinson, J. P. (2012). Housework: Who did, does or will do it, and how much does it matter? *Social Forces, 91*(1), 55–63. https://doi.org/10.1093/sf/sos120

Burkinshaw, P., & White, K. (2017). Fixing the women or fixing universities: Women in HE leadership. *Administrative Sciences, 7*(3), 30–43. https://doi.org/10.3390/admsci7030030.

Canetto, S. S., Trott, C. D., Winterowd, E. M., Haruyama, D., & Johnson, A. (2017). Challenges to the choice discourse: Women's views of their family and academic-science career options and constraints. *Journal of Feminist Family Therapy, 29*(1–2), 4–27. https://doi.org/10.1080/08952833.2016.1273174

Dickson, M. (2019). Academic motherhood in the United Arab Emirates. *Gender, Place & Culture, 26*(5), 719–739. https://doi.org/10.1080/0966369X.2018.1555143

Dickson, M., & Tennant, L. (2019). Emirati University student mothers post-childbirth: Support systems in the home. *Journal of Women and Gender in Higher Education, 12*(1), 88–105. https://doi.org/10.1080/19407882.2018.1533479

Dickson, M., Midraj., J., Al Hakmani, R., McMinn, M., Elsori, D., Alhashmi, M. & Tedam, P. (2023). Academic Parenthood in the United Arab Emirates in the time of COVID-19. *Frontiers in Education*

Digest of Education Statistics. (2015). Full-time faculty in degree-granting postsecondary institutions, by race/ethnicity, sex, and academic rank: Fall 2009, fall 2011, and fall 2013. *National Center for Education Statistics.*

Engin, M., & McKeown, K. (2017). Motivation of Emirati males and females to study at higher education in the United Arab Emirates. *Journal of Further and Higher Education, 41*(5), 678–691. https://doi.org/10.1080/0309877X.2016.1159293

Findlow, S. (2007). Women, higher education and social transformation in the Arab Gulf. In C. Brock & Z. L. Levers (Eds.), *Aspects of education in the Middle East and Africa* (pp. 57–76). Symposium Books.

Fitzgerald, T. (2013). *Women leaders in higher education: Shattering the myths*. Routledge.

Fothergill, A., & Feltey, K. (2003). "I've worked very hard and slept very little": Mothers on tenure track in academia. *Journal of the Motherhood Initiative for Research and Community Involvement, 5*(2).

Guarino, C. M., & Borden, V. M. (2017). Faculty service loads and gender: Are women taking care of the academic family? *Research in Higher Education, 58*(6), 672–694. https://doi.org/10.1007/s11162-017-9454-2

Heijstra, T., Bjarnason, T., & Rafnsdóttir, G. L. (2015). Predictors of gender inequalities in the rank of full professor. *Scandinavian Journal of Educational Research, 59*(2), 214–230. https://doi.org/10.1080/00313831.2014.904417

Heijstra, T. M., Steinthorsdóttir, F. S., & Einarsdóttir, T. (2017). Academic career making and the double-edged role of academic housework. *Gender and Education, 29*(6), 764–780. https://doi.org/10.1080/09540253.2016.1171825

Huang, C. (2013). Gender differences in academic self-efficacy: A meta-analysis. *European Journal of Psychology of Education, 28*(1), 1–35. https://doi.org/10.1007/s10212-011-0097-y

Kalm, S. (2019). On academic housekeeping and its allocation. *Sociologisk Forskning, 56*(1), 5–26.

Knobloch-Westerwick, S., Glynn, C. J., & Huge, M. (2013). The Matilda effect in science communication: An experiment on gender bias in publication quality perceptions and collaboration interest. *Science Communication, 35*(5), 603–625. https://doi.org/10.1177/1075547012472684

Macfarlane, B., & Burg, D. (2019). Women professors and the academic housework trap. *Journal of Higher Education Policy and Management, 41*(3), 262–274. https://doi.org/10.1080/1360080X.2019.1589682

Madsen, S. R., & Cook, B. J. (2010). Transformative learning: UAE, women, and higher education. *Journal of Global Responsibility, 1*(1), 127–148. https://doi.org/10.1108/20412561011039744

Melkonian, M., & Areepattamannil, S. (2018). The effect of absolute age-position on academic performance: A study of secondary students in the United Arab Emirates. *Educational Studies, 44*(5), 551–563. https://doi.org/10.1080/03055698.2017.1382330

Miller, K. E. (2021). The ethics of care and academic motherhood amid COVID-19. *Gender, Work & Organization, 28*, 260–265. https://doi.org/10.1111/gwao.12547

Misra, J., Lundquist, J. H., Holmes, E., & Agiomavritis, S. (2011). The ivory ceiling of service work. *Academe, 97*(1), 22.

Morley, L. (2013). The rules of the game: Women and the leaderist turn in higher education. *Gender and Education, 25*(1), 116–131. https://doi.org/10.1080/09540253.2012.740888

Morley, L. (2014). Lost leaders: Women in the global academy. *Higher Education Research & Development, 33*(1), 114–128. https://doi.org/10.1080/07294360.2013.864611

O'Connor, P. (2019). Gender imbalance in senior positions in higher education: What is the problem? What can be done? *Policy Reviews in Higher Education, 3*(1), 28–50. https://doi.org/10.1080/23322969.2018.1552084

Østby, G., Strand, H., Nordås, R., & Gleditsch, N. P. (2013). Gender gap or gender bias in peace research? Publication patterns and citation rates for Journal of Peace Research, 1983–2008. *International Studies Perspectives, 14*(4), 493–506. https://doi.org/10.1111/insp.12025

Perez, C. C. (2019). *Invisible women: Exposing data bias in a world designed for men*. Random House.

Reed, D. A., Enders, F., Lindor, R., McClees, M., & Lindor, K. D. (2011). Gender differences in academic productivity and leadership appointments of physicians throughout academic careers. *Academic Medicine, 86*(1), 43–47. https://doi.org/10.1097/ACM.0b013e3181ff9ff2

Salin, D. (2020). 'Competent' or 'considerate'? The persistence of gender bias in evaluation of leaders. *Nordic Journal of Working Life Studies, 10*(1), 59–79. https://doi.org/10.18291/njwls.v10i1.118680

Savigny, H. (2014). Women, know your limits: Cultural sexism in academia. *Gender and Education, 26*(7), 794–809. https://doi.org/10.1080/09540253.2014.970977

Schiebinger, L., & Gilmartin, S. K. (2010). Housework is an academic issue. *Academe, 96*(1), 39–44.

Schucan Bird, K. (2011). Do women publish fewer journal articles than men? Sex differences in publication productivity in the social sciences. *British Journal of Sociology of Education, 32*(6), 921–937. https://doi.org/10.1080/01425692.2011.596387

Shepherd, S. (2017). Why are there so few female leaders in higher education: A case of structure or agency? *Management in Education, 31*(2), 82–87. https://doi.org/10.1177/0892020617696631

Sim, W. (2020). For love, money and status, or personal growth? A survey of young Emirati women's educational aspirations. *Gulf Education and Social Policy Review (GESPR)*, 73–90. https://doi.org/10.18502/gespr.v1i1.7470.

Teele, D. L., & Thelen, K. (2017). Gender in the journals: Publication patterns in political science. *PS: Political Science & Politics, 50*(2), 433–447. https://doi.org/10.1017/S1049096516002985.

Van den Brink, M., & Benschop, Y. (2012). Slaying the seven-headed dragon: The quest for gender change in academia. *Gender, Work & Organization, 19*(1), 71–92. https://doi.org/10.1111/j.1468-0432.2011.00566.x

Wang, Y., Yue, Q., & Yu, K. (2013). Women leaders of higher education: Female executives in leading universities in China. *Cross-Cultural Communication, 9*(6), 40–45. https://doi.org/10.3968/j.ccc.1923670020130906.3031

Ward, K., & Wolf-Wendel, L. (2004). Academic motherhood: Managing complex roles in research universities. *The Review of Higher Education, 27*(2), 233–257. https://doi.org/10.1353/rhe.2003.0079.

West, J. D., Jacquet, J., King, M. M., Correll, S. J., & Bergstrom, C. T. (2013). The role of gender in scholarly authorship. *PloS one, 8*(7), e66212.

Winslow, S. (2010). Gender inequality and time allocations among academic faculty. *Gender & Society, 24*(6), 769–793. https://doi.org/10.1353/rhe.2003.0079